Cover design by Kristie Gianopulos
Front cover photo:
 Freshwater marsh at Historic Yates Millpond Preserve, Raleigh, NC
Back cover photos:
 (Left) Mountain bog at White Oak Bottom at Standing Indian Campground, Franklin, NC
 (Middle) Riverine forest wetland along Walnut Creek, Raleigh, NC
 (Right) Salt marsh at Cedar Point Tidelands Trail, Swansboro, NC

ISBN 978-0-578-86377-1

Guide to
Common Wetland Plants
of North Carolina

Kristie Gianopulos and Karen Kendig
with Milo Pyne

A publication by the North Carolina Division of Water Resources,
North Carolina Department of Environmental Quality
with funding from the US Environmental Protection Agency

Sagittaria latifolia

ACKNOWLEDGEMENTS

Roles of the Authors

Kristie Gianopulos is a wetland scientist with the NC Department of Environmental Quality. She compiled all information in this guide, took thousands of photographs, created the graphic design and layout of this guide, created supplemental pages, and provided some drawings.

Karen M. Kendig (formerly Karen M. Lynch) is a retired biologist from the NC Department of Environmental Quality and NC Department of Transportation. She authored and illustrated the original 1997 Common Wetland Plants of North Carolina guide. For this updated revision, she provided additional illustrations and input on plant species.

Milo Pyne is an ecologist and botanist, retired from NatureServe and NC Natural Heritage Program. He assisted Kristie Gianopulos in locating and identifying plants for photographs, provided photographs, and gave input on all species descriptions and the glossary.

Photographs/Illustrations

Note: Although this guide was published using federal and state funds, all illustrations were used by permission. Reuse of illustrations herein requires permission from the original artists or copyright holders.

All photographs were taken by Kristie Gianopulos, unless otherwise noted. Drawings were used by permission from a number of sources, in addition to the original drawings by Karen Kendig. Additional drawings were used by permission from USDA PLANTS website and Southeastern field office, the Flora of North America Association, the Manual of Marsh and Aquatic Vascular Plants of North Carolina with Habitat Data by Ernest O. Beal, illustrated by Sara Fish Brown (1977), and the Manual of the Vascular Flora of the Carolinas by Radford, Ahles, and Bell (1968). Kristie Gianopulos provided line drawing of *Commelina virginica*, *Galium tinctorium*, and *Smilax glauca*. Line drawing illustrations between sections were obtained from various sources copyright free with permission.

Other Acknowledgements

Many thanks to those who granted site access or tours to gather photos and plant information across North Carolina (NC Wildlife Resources Commission, US Forest Service, and The Nature Conservancy). Dr. Tom Wentworth assisted in finding contacts for the E.O. Beal guide illustrations and Dr. Alan Weakley granted permission to use illustrations from Radford et al. (1968). Kristie Gianopulos, Greg Rubino, and Amanda Mueller co-wrote the text on wetland types; Amanda Mueller provided many photographs and assisted with invaluable editorial review of this edition.

Arisaema triphyllum

CONTENTS

ABOUT THE GUIDE........1

GENERAL WETLAND TYPES IN NORTH CAROLINA........5

PLANT TERMINOLOGY
ILLUSTRATED......12
GLOSSARY......16

TREES......21

SHRUBS......87

FERNS....165

MONOCOT HERBS....179

DICOT HERBS....281

VINES....357

AQUATICS....385

COMMON CONFUSIONS....401

INDEX....412

Saururus cernuus

ABOUT THE GUIDE

Wetlands are valuable, vanishing resources that provide useful functions including water storage and purification, wildlife and aquatic habitat, and outdoor recreation and education. We hope that visitors to wetlands will recognize and appreciate the value of these wonderlands, beginning with the observation of wetland plants and animals.

Wetland plants are indicators of the duration of water in an area. The ability to identify wetland plants is useful to anyone who needs to know where a wetland lies, whether they be landowner, regulator, developer, government, or conservation organization. This guide provides identifying information for over 200 species of wetland plants commonly found in North Carolina's diverse wetlands.

This version is an update to the 1997 Common Wetland Plants of North Carolina written and illustrated by Karen Kendig before she retired from NC DEQ (then NC DENR). The original guide included 128 taxa; this newly revised edition expands the guide to 206 taxa and now includes photographs of all taxa, notation of non-native species, a section on commonly confused species, and other updated information.

Plant Species Included in This Guide

Gauging Commonness

We relied heavily on abundance and distribution information from the Vascular Plants of North Carolina biodiversity project website (LeGrand et al. 2020) which compiles information from botanical collections, NC Natural Heritage Program data, Dr. Weakley's flora (Weakley 2020), and extensive professional field experience of the website authors. Additionally, we ensured that our guide included the most geographically widespread wetland species in the state, as well as dominant species from wetland plots in the Carolina Vegetation Survey and NC Division of Water Resources vegetation plot surveys in wetlands across the state.

Wetland Status

Because plants have different tolerances for hydric conditions, wetland plant indicator statuses have been assigned to plant species for each region of the United States, and listed in the National Wetland Plant List (NWPL) by the US Army Corps of Engineers. Common North Carolina species were included if they had a status of facultative (FAC), facultative wetland (FACW), or obligate (OBL) for one of the state's ecoregions in 2021 (Figure 1; Table 1). Each species' current wetland indicator status is reported in this guide by ecoregion.

Figure 1. National Wetland Plant List Regions in North Carolina

Mountains and Piedmont

Coastal Plain

Table 1. National Wetland Plant List Indicator Status

Obligate Wetland (OBL)	plants that are essentially restricted to wetlands
Facultative Wetland (FACW)	plants that usually occur in wetlands, but may occur elsewhere
Facultative (FAC)	plants that occur in wetlands and non-wetlands
Facultative Upland (FACU)	plants that usually occur in non-wetlands, but may occur in wetlands
Upland (UPL)	plants that are essentially restricted to uplands

Notes on Range Maps

Species ecoregion range maps on each page were generated based on relative commonness levels determined by Weakley's (2020) and LeGrand et al.'s (2020) information on relative plant abundance (Figure 2). The range maps provided are intended to be general guides as to where species are most commonly found. There will be instances where a species is found primarily in one ecoregion and in a few counties in another ecoregion. In these cases, the range map provided shows the ecoregion(s) of primary occurrence, with notes on additional ecoregion(s) with occasional occurrence outside the one(s) shown.

Coefficient of Conservatism Ratings

Each species page in this guide includes the Coefficient of Conservatism assigned to that species, along with its wetland indicator status. Coefficients of Conservatism are a system of rating individual species for their habitat affinity, whether they be restricted to pristine areas or are tolerant of areas disturbed by human activities. Each species has been assigned an ecoregion-based rating on a scale of 0 to 10 by a panel of expert field botanists, where 10 represents species extremely intolerant of human-caused disturbance, and 1 represents species found nearly always in highly disturbed areas (Taft et al. 1997; Gianopulos 2014). Non-native species receive a rating of 0. These coefficients can be used to calculate floristic quality index values for an area based on species composition, and indicate relative quality of plant communities. The complete database of these values for wetland plants of the Southeast is available for download at www.ncwetlands.org/research and elsewhere.

Notes on Nomenclature

Scientific names in this guide are from the USDA PLANTS database as of July 1, 2020; this the nomenclature standard used in the NWPL and the NC Division of Water Resources' Coefficient of Conservatism value database. For the few cases where there was disagreement between USDA nomenclature and the NWPL, we gave preference to the (newer) name on the NWPL (these included *Persicaria sagittata*, *Osmundastrum cinnamomeum*, and *Osmunda spectabilis*). We discovered six instances where family classifications conflicted between USDA PLANTS and Weakley (2020), and we chose to follow Weakley: *Hypoxis hirsuta* (Hypoxidaceae rather than Liliaceae), *Mimulus* spp. (Phrymaceae rather than Scrophulariaceae), *Penthorum sedoides* (Penthoraceae rather than Crassulaceae), *Sambucus nigra* and *Viburnum* spp. (Adoxaceae rather than Caprifoliaceae), and *Zephyranthes atamasca* (Amaryllidaceae rather than Liliaceae).

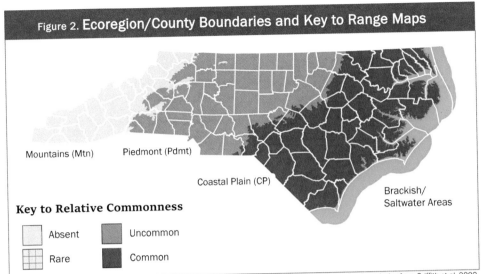

Figure 2. Ecoregion/County Boundaries and Key to Range Maps

Ecoregion boundaries from Griffith et al. 2002

Resources and Illustration Sources

Beal, E.O. 1977. A Manual of Marsh and Aquatic Vascular Plants of North Carolina with Habitat Data. Illustrated by Sara Fish Brown. NC Agricultural Research Service, Raleigh, North Carolina.

Britton, N.L., and A. Brown. 1913. An illustrated flora of the northern United States, Canada and the British Possessions. 3 vols. Charles Scribner's Sons, New York.

Flora of North America Editorial Committee, eds. 1993+. Flora of North America North of Mexico [Online]. 21+ vols. New York and Oxford. Vol. 1, 1993; vol. 2, 1993; vol. 3, 1997; vol. 4, 2003; vol. 5, 2005; vol. 7, 2010; vol. 8, 2009; vol. 19, 2006; vol. 20, 2006; vol. 21, 2006; vol. 22, 2000; vol. 23, 2002; vol. 24, 2007; vol. 25, 2003; vol. 26, 2002; vol. 27, 2007; vol 28, 2014; vol. 9, 2014; vol. 6, 2015; vol. 12, 2016; vol. 17, 2019. http://floranorthamerica.org. Accessed April 2, 2021.

Gianopulos, K. 2014. Coefficient of Conservatism Database Development for Wetland Plants Occurring in the Southeast United States. NC Dept. of Environmental Quality, Division of Water Resources. Report to the EPA, Region 4. 87 pp.

Griffith, G.E., Omernik, J.M., Comstock, J.A., Schafale, M.P., McNab, W.H., Lenat, D.R., MacPherson, T.F., Glover, J.B., and Shelburne, V.B. 2002. Ecoregions of North Carolina and South Carolina: Reston, Virginia. U.S. Geological Survey (map scale 1:1,500,000).

LeGrand, H., B. Sorrie, and T. Howard. 2020. *Vascular Plants of North Carolina* [Online]. Raleigh, NC : North Carolina Biodiversity Project and North Carolina State Parks. Available from https://auth1.dpr.ncparks.gov/flora/index.php

Radford, A., H. Ahles, C.R. Bell. 1968. Manual of the Vascular Flora of the Carolinas. University of North Carolina Press. 1183 pp.

Taft, J.B., Wilhelm, G.S., Ladd, D.M. & Masters, L.A. 1997. Floristic quality assessment for vegetation in Illinois, a method for assessing vegetation integrity. Illinois Native Plant Society, Westville, IL.

Weakley, A. 2020. Flora of the southern and mid-Atlantic states. University of North Carolina at Chapel Hill Herbarium, North Carolina Botanical Garden, University of North Carolina at Chapel Hill, Chapel Hill, NC. Available from https://ncbg.unc.edu/research/unc-herbarium/floras/

GENERAL WETLANDS TYPES IN NORTH CAROLINA

North Carolina has many kinds of natural systems ranging from ancient mountains to barrier island beaches. We have shrubby bogs in the mountains, lowland wetland forests in the Piedmont, freshwater marshes and Carolina bays in the flat Coastal Plain, and brackish marshes thriving behind barrier islands at the coast. Our amazing diversity of wetlands means we have a wide variety of plants commonly found in our state. Correctly identifying plant species within a wetland can increase your understanding and appreciation for the type of wetland you are in.

The following categories are major wetland types found within North Carolina. Other variants and subtypes of these exist in more restricted extent, an example being maritime forest wetlands which occur only on our barrier islands. Additionally, various organizations divide North Carolina's wetlands into more categories, such as the 16 wetland types used by the NC Wetland Assessment Method for permitting purposes, and many more wetland types are described in the classification used by the NC Natural Heritage Program. The major wetland types listed here are a simplification of these many classifications.

Mountain Bog

Mountain bogs in North Carolina are wetlands found in relatively flat spaces, usually at the bases of slopes, where long-term ground saturation creates an ideal situation for a range of wetland plants, many of them unique non-woody plants. Bogs are often quite small; many are only one acre or less. Thick deposits of organic soils allow threatened and endangered plants like mountain sweet pitcher plant (*Sarracenia jonesii*), green pitcher plant (*Sarracenia oreophil*), swamp pink (*Helonias bullata*), and bunched arrowhead (*Sagittaria fasciculata*) to grow, as well as unique wildlife like the critically endangered bog turtle (*Glyptemys muhlenbergii*). Bogs also support extensive mats of sphagnum moss, in accumulations not found in other natural landscapes. Walking in a bog feels like walking or bouncing on a wet sponge. Only 20% of North Carolina bogs that existed prior to European settlement still exist in their unaltered condition because many have been ditched or filled for farming.

> → *Common dominant plant species in mountain bogs:*
> Red maple (*Acer rubrum*), tag alder (*Alnus serrulata*), swamp rose (*Rosa palustris*), sphagnum moss (*Sphagnum* spp.), rushes (*Juncus* spp.), bulrushes (*Scirpus* spp.), sedges (*Carex* spp., *Cyperus* spp.), cinnamon fern (*Osmundastrum cinnamomeum*), royal fern (*Osmunda spectabilis*), sensitive fern (*Onoclea sensibilis*)

Riverine Forest Wetland (Brownwater and Blackwater)

Riverine forest wetlands occur extensively in North Carolina, other southern states, and even in other countries. They occur throughout North Carolina along river floodplains, receiving much of their water from floods, but they also receive water from rain, surface runoff, and/or groundwater at or near the surface. These rivers can be brownwater (carrying sediment and originating from the Piedmont or Mountains) or blackwater (not carrying sediment; tannic in nature and flowing through peat-based areas in the Coastal Plain).

Bottomland hardwood forests and riverine swamp forests are two riverine forest wetland types that often occur in close association with each other and can be difficult to tell apart. Riverine swamp forests may transition into bottomland hardwood forests further away from a river. Bottomland hardwood forests occur along second-order or larger streams and provide good water storage, especially during periods of high rains. They may contain floodplain pools and may transition upstream into forested headwater wetlands. The size and vigor of these valuable wetland forests are tied to the condition of the river and its floodplain, so changing these systems can significantly alter the services provided by these wetlands.

Many species of trees found in riverine swamp forests and bottomland hardwood forests provide valuable lumber and many of these trees are buttressed to keep them stable in soft soils. Water-tolerant trees like red maple (*Acer rubrum*), overcup oak (*Quercus lyrata*), blackgum (*Nyssa sylvatica*), ashes (*Fraxinus* spp.), American elm (*Ulmus americana*), cypress (*Taxodium* spp.), and Atlantic white cedar (*Chamaecyparis thyoides*) constitute the majority of the plant cover, though shrubs and non-woody plants are also found in these wetlands. Beavers are a major contributing factor in swamp forest creation. These swamp forests are important places for maintaining plant diversity and providing habitat for many birds, fish, invertebrates, and mammals.

Large swaths of riverine swamp forests are found in the Coastal Plain, but significant tracts are also found in the Piedmont, especially adjacent to large lakes and in low-lying areas along large rivers that can be flooded most, if not all, of the year with typical rain levels.

→ *Common dominant plant species in riverine forest wetlands:*
   Mountains:

   Red maple (*Acer rubrum*), American sycamore (*Platanus occidentalis*), river birch (*Betula nigra*), tag alder (*Alnus serrulata*), royal fern (*Osmunda spectabilis*), jewelweed (*Impatiens capensis*), lizard's tail (*Saururus cernuus*), sensitive fern (*Onoclea sensibilis*)

Piedmont and Coastal Plain:

Bald cypress (*Taxodium distichum*), water tupelo (*Nyssa aquatica*), swamp tupelo (*Nyssa biflora*), red maple (*Acer rubrum*), willow oak (*Quercus phellos*), green ash (*Fraxinus pennsylvanica*), American elm (*Ulmus americana*), sweetgum (*Liquidambar styraciflua*), river birch (*Betula nigra*), ironwood (*Carpinus caroliniana*), swamp chestnut oak (*Quercus michauxii*), water oak (*Quercus nigra*), eastern poison ivy (*Toxicodendron radicans*), greenbrier (*Smilax* spp.), painted buckeye (*Aesculus sylvatica*), coastal dog hobble (*Leucothoe axillaris*), swamp fetterbush (*Eubotrys racemosus*), Virginia sweetspire (*Itea virginica*), coastal sweet-pepperbush (*Clethra alnifolia*), swamp titi (*Cyrilla racemiflora*), lizard's tail (*Saururus cernuus*), jewelweed (*Impatiens capensis*), giant cane (*Arundinaria gigantea*), cinnamon fern (*Osmundastrum cinnamomeum*), royal fern (*Osmunda spectabilis*), sensitive fern (*Onoclea sensibilis*), Virginia chain fern (*Woodwardia virginica*), netted chain fern (*Woodwardia areolata*), Canadian clearweed (*Pilea pumila*), false nettle (*Boehmeria cylindrica*)

Seep Wetland

Seep wetlands are so named because groundwater "seeps" to the surface. These wetlands are typically located along slopes leading down to floodplains. Seep wetlands are characterized by the presence of slow-moving groundwater that is unable to permeate the underlying clay or bedrock and therefore is being discharged to the surface. The soil is saturated for most, if not all, of the year, but there is usually not enough water to flow and form channels in a seep. Vegetation can range from small non-woody plants to large hardwood trees, and wildlife is varied as well. Salamanders frequently find refuge in seep wetlands, especially in the mountains.

→ *Common dominant plant species in seep wetlands*:

Mountains and Piedmont:

Red maple (*Acer rubrum*), sweetgum (*Liquidambar styraciflua*), green ash (*Fraxinus pennsylvanica*), willow oak (*Quercus phellos*), possumhaw viburnum (*Viburnum nudum*), cinnamon fern (*Osmundastrum cinnamomeum*), royal fern (*Osmunda spectabilis*), jewelweed (*Impatiens capensis*), false nettle (*Boehmeria cylindrica*)

Coastal Plain:

Sweetbay (*Magnolia virginiana*), sweetgum (*Liquidambar styraciflua*), red maple (*Acer rubrum*), sweet-pepperbush (*Clethra alnifolia*), fetterbush (*Lyonia lucida*), cinnamon fern (*Osmundastrum cinnamomeum*)

Freshwater Marsh

Freshwater marshes can be found in a wide variety of wet landscape contexts including along rivers, canals, and estuaries, along lakes or ponds, along the base of wet slopes, and in wet meadows (relatively flat areas where the groundwater table is naturally close to the ground surface). Marshes generally do not have trees. They have a wide variety of small, non-woody, grass-like plants, and soft-tissued flowering plants with broad leaves. Many times, freshwater marshes are a result of disturbances, such as forest fire, utility line maintenance, or beaver activity changing streams to ponded areas that slowly fill with sediment. The deep organic and mineral soils are continuously saturated, if not flooded, enabling thick deposits of plant material to accumulate. Marshes are enormously important to our state, helping to limit flood damage during storms and hosting important wildlife including birds, mammals, and commercial fish species.

In some cases, freshwater marshes can be tidally influenced, when far enough up in the watershed to be freshwater but still experience fluctuations in water level from nearby streams or rivers. Tidal freshwater marshes are a type of freshwater marsh found only in the Coastal Plain, in areas where they experience flooding by lunar and wind tides and during temporary high wind storms. Salinity is typically low (< 0.5 parts per thousand), and these marshes support a large diversity of herbaceous and some woody plants. Tidal freshwater marshes do not comprise a large portion of freshwater marshes in North Carolina, compared to other southeastern states, because most of the state's rivers drain to large sounds, rather than directly to the ocean, and tidal influence upstream is reduced as a result.

→ *Common dominant plant species in freshwater marshes*:
Cattail (*Typha* spp.), rushes (*Juncus* spp.), sedges (*Carex* spp.; *Cyperus* spp.), beakrushes (*Rhynchospora* spp.), bulrushes (*Scirpus* spp.), spikerushes (*Eleocharis* spp.), black willow (*Salix nigra*), common buttonbush (*Cephalanthus occidentalis*), tag alder (*Alnus serrulata*), swamp rose (*Rosa palustris*), common wax myrtle (*Myrica cerifera*), swamp rose mallow (*Hibiscus moscheutos*), green arrow arum (*Peltandra virginica*), Virginia iris (*Iris virginica*), bladderworts (*Utricularia* spp.), duckweed (*Lemna* spp.), yellow pond-lily (*Nuphar lutea*), American water-lily (*Nymphaea odorata*), American lotus (*Nelumbo lutea*)

Pocosin

Pocosins are unique to the southeastern portion of the Atlantic Coastal Plain with the majority of them occurring in North Carolina. The word "pocosin" is an Algonquian Native American term meaning "swamp on a hill." Pocosins develop in areas where ancient river valleys were filled in with sand, silt, and clay from the eroding Appalachian Mountains to the west. Over thousands of years, dead plant matter accumulated in these wet, low places, building up in slightly hilly formations. Because of their landscape position, these areas do not naturally have streams flowing into them.

Pocosins can be very demanding places for plants to grow. The water is acidic, low in nutrients, and only replenished by precipitation. Very few kinds of trees can survive in a true pocosin, one of them being the gnarly pond pine (*Pinus serotina*). Evergreen shrubs grow very densely in pocosins and include a variety of bays (*Magnolia virginiana, Gordonia lasianthus, Persea borbonia*), hollies (*Ilex* spp.), and greenbriers (*Smilax* spp.). Very few of the grass-like plant species or non-woody plants we are used to seeing in wetlands are found in pocosins, and pocosins can be very difficult to walk through because they are so dense. Pocosins once covered much of the eastern third of our state, but many of them have been converted to agricultural land by removing water through ditching.

→ *Common dominant plant species in pocosins:*
   Pond pine (*Pinus serotina*), sweetbay (*Magnolia virginiana*), loblolly bay (*Gordonia lasianthus*), swamp bay (*Persea borbonia*), fetterbush lyonia (*Lyonia lucida*), swamp titi (*Cyrilla racemiflora*), hollies (*Ilex* spp.), laurel greenbrier (*Smilax laurifolia*)

Pine Wetland

Pine wetlands (pine savannas or pine flats) are a type of forested wetland found in sandy soils of the Coastal Plain states, including North Carolina. Water table levels in these wetlands stay close enough to the ground surface to foster a variety of water-tolerating plants, including trees, shrubs, and herbs. Plants that can survive in these conditions include bays (*Magnolia virginiana, Gordonia lasianthus, Persea borbonia*), fetterbush lyonia (*Lyonia lucida*), inkberry (*Ilex glabra*), common wax myrtle (*Morella cerifera*), southern bayberry (*Morella caroliniensis*), beakrushes (*Rhynchospora* spp.), orchids, and lilies. Pine savannas are dominated by longleaf pines (*Pinus palustris*) standing over a groundcover of grasses and other herbaceous plants. This plant community

is dependent on cycles of destruction and regrowth from forest fires. This natural influence keeps non-native species out and allows a wonderful diversity of native species to thrive, including rare species like the Venus fly-trap (*Dionaea muscipula*). The valuable longleaf pines have been extensively logged in North Carolina, drastically reducing the area of natural savannas to about 5% of their original extent. Many altered pine flat wetlands can still be found as successors of previous savannas, swamp forests, or active pine plantations. Trees in today's typical pine flat wetland tend to be loblolly pines (*Pinus taeda*) and slash pines (*Pinus elliottii*), though other pines and hardwoods like maples (*Acer* spp.) may be found in them as well. Natural pine wetlands in North Carolina harbor many interesting rare animals including the red-cockaded woodpecker, pine barrens treefrog, pine snake, gopher frog, and fox squirrel.

> → *Common dominant plant species in pine wetlands:*
> Loblolly pine (*Pinus taeda*), red maple (*Acer rubrum*), sweetgum (*Liquidambar styraciflua*), swamp tupelo (*Nyssa biflora*), longleaf pine (*Pinus palustris*), pond pine (*Pinus serotina*), common sweetleaf (*Symplocos tinctoria*), hollies (*Ilex* spp.), giant cane (*Arundinaria gigantea*), wiregrass (*Aristida stricta*), beakrushes (*Rhynchospora* spp.), southern bayberry (*Morella caroliniensis*)

Salt Marsh and Brackish Marsh

Salty ocean tides provide regular flooding of salt marshes and brackish marshes. Salt marshes (salinity greater than 30 ppt) contain species like saltmeadow cordgrass (*Spartina patens*) and smooth cordgrass (*Spartina alterniflora*). If a marsh receives a mix of freshwater and saltwater, it will be brackish (salinity of 0.5 to 30 parts per thousand). Brackish marshes are usually dominated by black needlerush (*Juncus roemerianus*). Both types of coastal marsh are important nursery grounds for commercially important fish and shellfish. They also help absorb wave energy during storms.

> → *Common dominant plant species in salt marshes:*
> Smooth cordgrass (*Spartina alterniflora*), saltmeadow cordgrass (*Spartina patens*), saltgrass (*Distichlis spicata*), glasswort (*Salicornia* spp.)

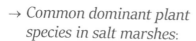

> → *Common dominant plant species in brackish marshes:*
> Black needlerush (*Juncus roemerianus*), big cordgrass (*Spartina cynosuroides*),

swamp sawgrass (*Cladium mariscus*), saltmeadow cordgrass (*Spartina patens*), saltgrass (*Distichlis spicata*)

Estuarine Woody Wetland

Estuarine woody wetlands are coastal, found at the edges of estuaries or salt marshes. Their water levels are a bit unpredictable since they are affected by wind tides and occasional flooding from salt or brackish water. The plant life found in these wetland types is primarily woody vegetation including trees and large shrubs. These trees and shrubs are able to adapt to changing water levels and chemical influences from saltwater.

→ *Common dominant plant species in estuarine woody wetlands:*
Marsh elder (*Iva frutescens*), eastern baccharis (*Baccharis halimifolia*), common wax myrtle (*Myrica cerifera*), Atlantic white cedar (*Chamaecyparis thyoides*), sea ox-eye daisy (*Borrichia frutescens*), swamp sawgrass (*Cladium mariscus*), saltmeadow cordgrass (*Spartina patens*), saltgrass (*Distichlis spicata*)

Carolina Bay

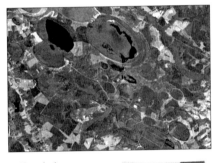

Carolina bays are not a wetland type per se, but they are unique, elliptical landscape features found in the Coastal Plain. There are many in North Carolina and, along with open water, they can also contain a variety of wetland types (e.g., pocosins, freshwater marsh, cypress swamp). The name "Carolina Bay" comes not from the location of these wetlands in relation to open water, but from the consistent presence of bay trees (loblolly bay, swamp bay, and sweetbay). These large and small oval depressions in the landscape are not associated with streams or other water bodies, and they are all oriented in the same northwestern direction. Geological experts agree they were created by scooping action of glacial winds over semi-arid plains south of the glacial ice. They were filled in with water and wetland plants as sea level rose with glacial retreat. The plants, many of which are rare or endangered, benefit from the deep, rich soil in Carolina bays. The primary source of water in Carolina bays is precipitation during winter and spring months. In the summer, the water in shallower bays often dries up, making these wetlands excellent habitat for amphibians in particular. The temporary nature of the water reduces predators on eggs and tadpoles. A large number of these wetland treasures have been severely altered and/or converted by ditching and draining.

PLANT TERMINOLOGY (Illustrated)

Plant Parts

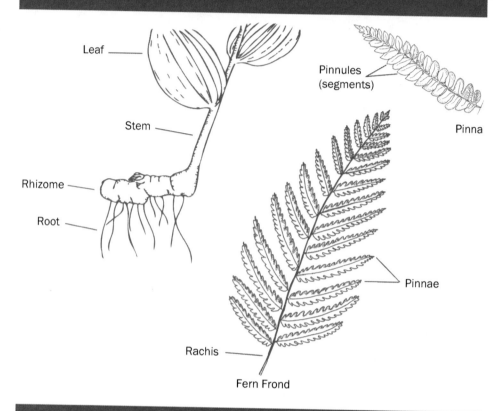

Flower Parts

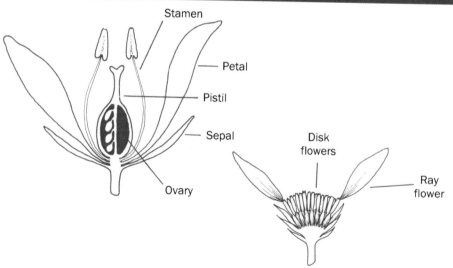

Flower Types

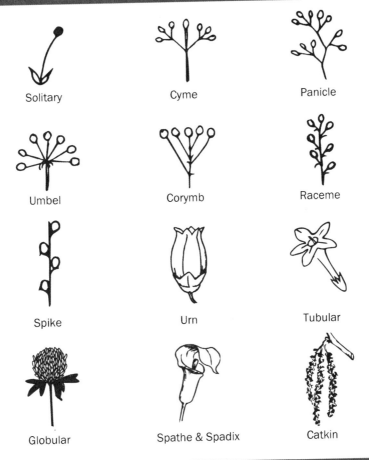

Fruit Types (selected)

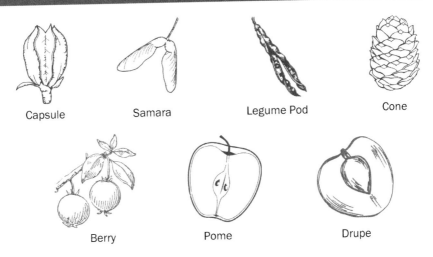

Leaf Shapes

 Elliptic

 Palmate

 Cordate

 Lanceolate

 Linear

 Oblanceolate

 Ovate

 Sagittate

 Obovate

 Hastate

 Scale

 Needle

Leaf Tips

 Acute

 Acuminate

 Rounded

Leaf Margins

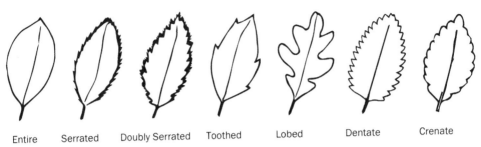

Entire Serrated Doubly Serrated Toothed Lobed Dentate Crenate

Leaf Arrangements

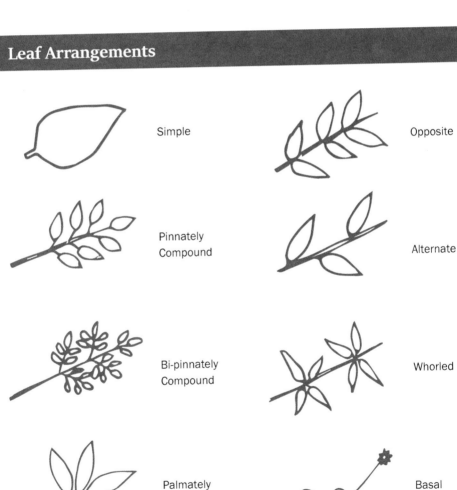

Simple

Opposite

Pinnately Compound

Alternate

Bi-pinnately Compound

Whorled

Palmately Compound

Basal Rosette

PLANT TERMINOLOGY (Glossary)

Achene - small, dry, one-seeded fruit with soft outer wall

Acuminate - with a long-tapering point

Acute - sharply pointed

Aerial root - root above the ground, not in contact with soil, sometimes anchoring the plant to a support

Alternate - leaves or leaflets arranged alternately along the axis (compare opposite, whorled)

Annual - plant whose life cycle is completed in one year or one season

Apex - the pointed end or tip (of leaf)

Appressed - lying flat against or nearly parallel to the surface (of a stem or leaf)

Aquatic - growing only in water

Aromatic - fragrant, especially when crushed

Awn - stiff, bristle-shaped appendage, usually at the end of a structure

Barb - hook-like projection

Basal - at the base or at ground level

Bi-pinnately compound/divided - compound leaf with leaflets divided into subleaflets

Blade - leaf like a grass, or the widest part of a leaf, distinct from a petiole

Bract - modified leaf which may be reduced in size or different from the foliage leaves and which usually subtends a flower or an inflorescence

Bristle - stiff hair-like structure, standing upright away from stem or seed

Bundle scar - mark from vascular bundles left on leaf scar when leaf falls

Buttressed - swollen and spreading trunk at the base of a tree, to aid in stabilization

Capsule - dry fruit, with rows of seeds and splitting in compartments at maturity

Carnivorous - meat-eating, or insect-eating as in the case of carnivorous plants

Catkin - a spikelike, often drooping, inflorescence with small flowers subtended by scales

Chambered pith - when the pith or inner spongy portion of branches is divided into sections or rooms

Clasping - having the lower edges of a leaf blade partly surrounding the stem

Compound - divided into two or more separate parts (leaflets)

Conical - shaped like a cone

Cordate - shaped like a heart

Corolla - all petals of a flower, whether fused or separated

Corymb - flat-topped inflorescence in which individual flowering stems arise from various points on stem to the same height; outermost flowers bloom first. [e.g. *Lachnanthes caroliniana* (redroot)]

Crenate - leaf margin with shallow, rounded teeth or lobes; wavy-edged

Culm - stem of a grass or grass-like plant

Cyme - a broad, flat inflorescence with a main center stem topped with a single flower; innermost flowers bloom first. Example: *Rhynchospora* spp.

Deciduous - plants that lose their leaves once a year, usually in winter

Dentate - toothed or jagged with sharp points perpedicular or at an angle to the leaf margin

Diaphragmed - when the pith or inner spongy portion of branches contains membranous divisions or lines

Disk and ray - flower comprised of combination of small, tubular flowers in the center, and outer ray flowers, each with a linear petal extension of the corolla

Doubly serrated - leaf margin where each sharp tooth is divided into smaller sharp teeth

Drupe - a fleshy, berry-like fruit containing one seed

Elliptic - football-shaped, with the widest point at the middle

Entire - smooth leaf margins, without teeth or divisions

Evergreen - plants which retain their leaves throughout the year

Frond - the entire leaf of a fern including the stipe (petiole), rachis (stem), and leaf blade

Furrow - long, deep groove, as in bark

Glabrous - smooth, without hairs

Glaucous - gray-green in color, or having a thin, whitish, powdery coating like that on grapes

Globose - spherical, globe-shaped

Graminoid - grass-like plant with elongated, linear leaf blades, usually referring to grasses, sedges, and rushes

Hastate - elongate, heart shaped with pointed basal lobes occurring at right angles to the petiole

Holdfast - aerial root that functions to anchor a climbing plant to its support

Inflorescence - flowering portion of a plant (a group of flowers)

Lanceolate - lance-shaped or spear-shaped; much longer than wide, with the broader end near the base

Leaflet - each of the separate leaf-like structures that make up a compound leaf

Lenticel - enlarged pore that functions in gas exchange, appears as a dot on stem

Ligule - a strap-shaped individual flower in a head in the Asteraceae family that contains only ray flowers, or ligules (e.g. dandelion)

Lobed - rounded extensions along a leaf margin; rounded sections of a fruit, divided by a line or split

Margin - outer edge of a leaf

Midrib - center vein in a leaf, usually extending from petiole to tip

Node - area along a stem where leaves usually originate

Nutlet - small, dry, one-seeded fruit with a very thin or no outer wall

Oblanceolate - longer than wide, with the broader end near the apex

Obovate - oval in shape with the broader end above the middle

Opposite - leaves or leaflets born on stem in pairs from opposite sides

Ovate - oval in shape, with the broader end below the middle

Ovoid - egg-shaped

Palmate - leaf with several midrib veins radiating from a center point

Palmately compound - compound leaf with several (5 to 7) leaflets all attached at a center point

Panicle - irregularly branched inflorescence

Perennial - plant which grows year after year

Perigynium - sac-like structure enclosing the fruit in Carex spp.

Petiole - stalk supporting a leaf

Pinna/pinnae - leaflet or a division of the leaf blade (frond) in ferns

Pinnately compound - divided into leaflets arranged on either side of a stem or petiole, as in a feather

Pinnule - subdivisions of leaflets (pinnae) in a fern frond

Pith - soft center inside a stem or twig

Plume - long, soft, fluffy inflorescence

Pome - type of fleshy fruit with multiple small seeds in the center

Pubescent - hairy or fuzzy

Punctate - having a surface scattered with tiny, depressed dots

Raceme - spike-like inflorescence with the flowers stalked

Rachis - central stalk of a compound leaf, frond, or inflorescence

Resinous - exuding a sticky organic substance, insoluble in water

Rhizome - elongated, underground root, which usually grows horizontally

Rosette - circular arrangement of leaves, attaching at a central point as in basal rosette (at ground level)

Sagittate - shaped like an arrowhead, triangular

Samara - flat, winged fruit; dispersed by wind; found on trees such as maples

Scabrous - having rough or finely toothed edges

Scale leaf - leaves that are thin, flat, and pressed closely together, as in some conifers like cedars

Scalloped - having shallow, rounded lobes

Semi-evergreen - plants that lose most, but not all, of their leaves for part of the year, usually winter

Sepal - modified leaves located at the base of the flower

Septate - partitioned into sections

Serrated/serrations - containing sharply pointed teeth

Sessile - attached by the base directly to the stalk without a petiole

Sheath - part of a leaf base that closely surrounds the main stem

Sori - clusters of spore cases on a fertile frond of a fern

Spadix - fleshy spike with flowers embedded along the axis

Spathe - leaf-like hood, open on one side, enclosing an inflorescence. Example: Jack-in-the-pulpit

Spherical - globe-shaped

Spike - elongated inflorescence with the flowers not stalked

Spikelet - unit of the inflorescence in the grass and sedge families, containing one or more flowers

Stamen - pollen bearing (male) part of a flower

Stipule - small, leaf-like outgrowth at base of leaf or petiole, often in pairs

Stipular scar - scar that is left on a twig when a leaf or petiole drops

Sub-opposite - leaf arrangement where leaves are not strictly alternate, but are not opposite

Succulent - containing fleshy tissue, not woody

Suture - rigid, linear division between two parts

Terminal - occurring at the tip

Terrestrial - occurring on land

Toothed - having triangular, tooth-like projections along the margin; encompasses the terms serrate, dentate, and crenate

Tuber - swollen, underground stem used for storage by certain plants

Tubular - similar to a hollow tube

Umbel - inflorescence in which flower stalks arise from a common point to a flat or convex top

Umbellet - subsection of an umbel with its own umbel arrangement; subumbel

Venation - pattern of veins in leaves

Whorled - leaf arrangement where three or more leaves emerge from each node

Liriodendron tulipifera

TREES

Scientific Name	Common Name	National Wetland Plant List Status Mtns & Piedmont/Coastal Plain	Page
Acer negundo	Boxelder	FAC/FAC	22
Acer rubrum	Red Maple	FAC/FAC	24
Betula nigra	River Birch	FACW/FACW	26
Carpinus caroliniana	Ironwood	FAC/FAC	28
Celtis laevigata	Sugarberry	FACW/FACW	30
Chamaecyparis thyoides	Atlantic White Cedar	OBL/OBL	32
Diospyros virginiana	Common Persimmon	FAC/FAC	34
Fraxinus caroliniana	Carolina Ash	OBL/OBL	36
Fraxinus pennsylvanica	Green Ash	FACW/FACW	38
Gordonia lasianthus	Loblolly Bay	(n/a)/FACW	40
Ilex opaca	American Holly	FACU/FAC	42
Liquidambar styraciflua	Sweetgum	FAC/FAC	44
Liriodendron tulipifera	Tuliptree	FACU/FACU	46
Magnolia virginiana	Sweetbay	FACW/FACW	48
Nyssa aquatica	Water Tupelo	OBL/OBL	50
Nyssa biflora	Swamp Tupelo	FACW/OBL	52
Nyssa sylvatica	Blackgum	FAC/FAC	54
Persea palustris	Swamp Bay	FACW/FACW	56
Pinus palustris	Longleaf Pine	FAC/FAC	58
Pinus serotina	Pond Pine	OBL/FACW	60
Pinus taeda	Loblolly Pine	FAC/FAC	62
Platanus occidentalis	American Sycamore	FACW/FACW	64
Quercus laurifolia	Laurel Oak	FACW/FACW	66
Quercus lyrata	Overcup Oak	OBL/OBL	68
Quercus michauxii	Swamp Chestnut Oak	FACW/FACW	70
Quercus nigra	Water Oak	FAC/FAC	72
Quercus pagoda	Cherrybark Oak	FACW/FAC	74
Quercus phellos	Willow Oak	FAC/FACW	76
Salix nigra	Black Willow	OBL/OBL	78
Taxodium ascendens	Pond Cypress	OBL/OBL	80
Taxodium distichum	Bald Cypress	OBL/OBL	82
Ulmus americana	American Elm	FACW/FAC	84

Acer negundo — Boxelder

Green, glossy twigs; opposite branching

Medium-sized tree

Toothed leaves in sets of 3 or 5

Winged, yellowish or tan samaras

Raised veins on leaf undersides

TREES

Aceraceae - Maple Family

Acer negundo
Boxelder

National Wetland Plant List
Mtn/Pdmt: **FAC**   CP: **FAC**

Coefficient of Conservatism
Mtn: **4**   Pdmt: **4**   CP: **3**

Drawing: Karen Kendig

Habit
Small to medium deciduous tree, reaching 25 m.

Leaves
Opposite, pinnately compound with an odd number (3-9) of leaflets, although 3 and 5 leaflets most common. Leaflets mostly ovate and toothed, and 5-10 cm long and 6 cm wide.

Flowers/Fruit
Fruits are paired light yellow or tan samaras, 3 cm long. Flowers March/April; fruits May to October.

Field Characteristics
Twigs glossy green with white lenticels.

Habitat/Range
Floodplains, stream banks, low woods of brownwater streams, throughout NC.

Similar Species
Sometimes confused with *Toxicodendron radicans* (eastern poison ivy). Look for more than 3 leaflets, distinctive green twigs, and opposite branching. Never a climbing vine like eastern poison ivy. See Common Confusions section p. 403.

Acer rubrum — Red Maple

Young trees and newer limbs have smooth bark

Winged samaras can be red or green

Covered with red buds and flowers in spring

Red flowers appear in early spring

Toothed leaves gray-green on back with red petioles

TREES

Aceraceae - Maple Family

Acer rubrum
Red Maple

National Wetland Plant List
Mtn/Pdmt: **FAC** CP: **FAC**

Coefficient of Conservatism
Mtn: **4** Pdmt: **3** CP: **3**

Drawing: Karen Kendig

Habit
Medium deciduous tree, sometimes reaching large stature.

Leaves
Opposite, lobed with teeth and with 3-5 main points. Leaves 6-14 cm long; can be as wide as long. Green above, lighter below.

Flowers/Fruit
Red flowers, January to March, before leafing out. Fruit is double samara, with each half about 3 cm long. Fruits February to July.

Field Characteristics
Opposite branching pattern. Showy clusters of reddish flowers in early spring.

Habitat/Range
Low woods, uplands, floodplains, swamps, stream banks, across NC. Very widespread in habitat.

Similar Species
Acer saccharum (sugar maple) in the mountains and *Acer floridanum* (Florida maple) in the Piedmont have similar, but untoothed, lobed leaves. *Platanus occidentalis* (American sycamore) has leaves of similar shape, but generally much larger, with bark that flakes off in patches.

Betula nigra — River Birch

Flaky, papery bark

Trees have male and female catkins

Doubly serrated edges on leaves

Female catkin

Triangular-shaped leaves

TREES

Betulaceae - Birch Family

Betula nigra
River Birch

National Wetland Plant List
Mtn/Pdmt: **FACW**   CP: **FACW**

Coefficient of Conservatism
Mtn: **5**   Pdmt: **4**   CP: **4**

Drawing: Karen Kendig

Habit
Deciduous, medium sized tree up to 25 m, with curly, peeling, papery bark.

Leaves
Alternate, doubly serrated, triangular or ovate leaves, 4-8 cm long. Leaves contain 7-9 straight veins on each side of leaf. Undersides lighter.

Flowers/Fruit
Male flowers drooping catkins and female flowers in a cone-like catkin. Fruits oblong samaras. Flowers March/April; fruits May/June.

Field Characteristics
Peeling bark and triangular leaves distinctive. Younger trees have rusty-colored bark, while older bark is less colorful and darker.

Habitat/Range
Floodplains, river and stream banks in moist soil. Found statewide, but chiefly Piedmont and Coastal Plain. Planted as an ornamental tree.

Similar Species
Leaves similar to *Carpinus caroliniana* (ironwood) and *Ostrya virginiana* (hop hornbeam), but bark is distinctive. See Common Confusions section p. 402.

Carpinus caroliniana — Ironwood

"Muscular" looking trunks

Three-lobed leafy bracted seeds in drooping clusters

Alternate, smooth leaves

Doubly serrated leaves; straight veins

Catkins appear at or before leaf emergence in spring

TREES

Betulaceae - Birch Family

National Wetland Plant List
Mtn/Pdmt: **FAC**   CP: **FAC**

Coefficient of Conservatism
Mtn: **5**   Pdmt: **5**   CP: **5**

Carpinus caroliniana
Ironwood

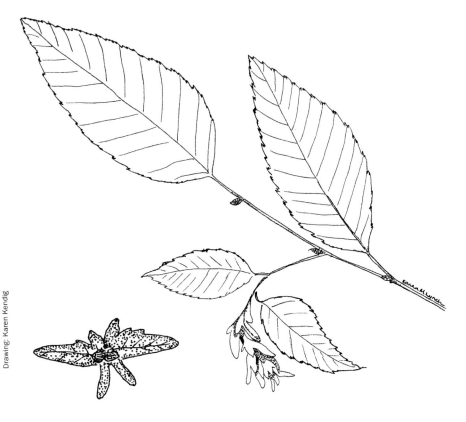

Drawing: Karen Kendig

Habit
Small, deciduous, understory tree with smooth, gray bark, up to 10 m.

Leaves
Alternate, ovate leaves, 3-15 cm long. Margins doubly serrated and leaves paler green and smooth on undersides. Leaf veins pronounced (particularly on leaf undersides) with straight veins running to the leaf edges.

Flowers/Fruit
Flowers in catkins: male catkin 3-4 cm long and female catkin about 2 cm long. Fruits are small nuts, subtended by a leafy 3-lobed bract in drooping clusters, about 10 cm long. Flowers March/April; fruits September/October.

Field Characteristics
Distinctive "muscular" branches and trunk.

Habitat/Range
Floodplain forests and bottomlands throughout NC.

Similar Species
Leaf shape similar to the pubescent leaves of *Ostrya virginiana* (hop hornbeam), but *Carpinus caroliniana* (ironwood) leaves are smooth and "muscular" looking trunks are distinctive. Can be confused with *Betula* spp. (birches); see Common Confusions section p. 402.

Celtis laevigata — Sugarberry

Bark with corky projections, or "warts"

Edible fruits August to October

Leaves are 3x longer than broad with prominent veins.

Simple, alternate leaves

Flowers appear early in spring

TREES

Ulmaceae - Elm Family

Celtis laevigata
Sugarberry

National Wetland Plant List
Mtn/Pdmt: **FACW**   CP: **FACW**
Coefficient of Conservatism
Mtn: **4**   Pdmt: **4**   CP: **4**

Habit
Medium to large tree, 25 to 30 m tall, with smooth, gray bark that has corky warts.

Leaves
Simple, alternate, lance-shaped leaves with uneven bases, prominent veins, and a length 3 times the width. Leaves glossy dark green above and yellowish-green below.

Flowers/Fruit
Edible fruits fleshy drupes with a large seed inside, deep reddish-purple when ripe. The taste is similar to dates. Flowers April/May; fruits August to October.

Field Characteristics
Look for corky, bumpy bark and long serrated leaves that are yellowish-green beneath. Leaves often covered with small galls.

Habitat/Range
Bottomland forests, natural levees, nearby upland forests, poorly drained clay. Uncommon in the Mountains and the Coastal Plain outside brownwater floodplains.

Chamaecyparis thyoides — Atlantic White Cedar

Medium sized evergreen tree

Reddish to gray-brown thin bark with flat, linear sections

Needles branch in one main, fan-like plane

Scale-like needles, not spined

Small, waxy cones; green immature, black-brown at maturity

TREES

Cupressaceae - Juniper Family

Chamaecyparis thyoides
Atlantic White Cedar

National Wetland Plant List
Mtn/Pdmt: **OBL**   CP: **OBL**

Coefficient of Conservatism
Mtn: **n/a**   Pdmt: **0**   CP: **9**

Not native in Piedmont

Drawing: Karen Kendig

Habit
Medium sized evergreen tree to 28 m in height.

Leaves
Flattened, scale-like leaves, 1-3 millimeters (mm) long and green on both sides.

Flowers/Fruit
Small inconspicuous cones; male cone is 2 mm long and female cone is spherical, 6 mm diameter, with a crumpled appearance. Blooms March/April; fruits October/November.

Field Characteristics
Evergreen needles are "flat". The wood is highly desirable, so this species was extensively logged in the past. Old-growth and extensive stands are now uncommon.

Habitat/Range
Acidic swamps of the Coastal Plain, generally in peaty soils or other poorly drained areas. Can grow in dense stands to the exclusion of other trees.

Similar Species
Juniperus virginiana (eastern red cedar) is a similar upland tree, with flattened needles branching in many planes instead of one main, fan-like plane as in *Chamaecyparis thyoides* (Atlantic white cedar).

Diospyros virginiana — Common Persimmon

Square-shaped blocks separated by furrows; gray bark

Smooth margins on leaves

Distinctive branching pattern with small buds

Twigs often purple; female flowers on short stems

Edible fruits ripen to orange-yellow

TREES

Ebenaceae - Ebony Family

National Wetland Plant List
Mtn/Pdmt: **FAC**   CP: **FAC**

Coefficient of Conservatism
Mtn: **4**   Pdmt: **4**   CP: **4**

Diospyros virginiana
Common Persimmon

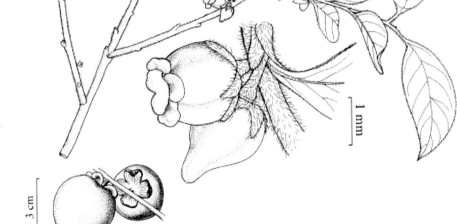

Drawing: Courtesy of the Flora of North America Association, Barbara Alongi

Habit
Small to medium deciduous tree up to 16 m.

Leaves
Alternate, ovate to elliptic leaves to 15 cm long, often with black splotches.

Flowers/Fruit
Male (yellow) and female (green), 4-petaled flowers occur on separate trees; petals are thick and waxy-looking. Male flowers grow in clusters of 2-3 and female flowers occur singly. Persimmon fruit is an orange berry, 2-4 cm wide. Flowers May/June; fruits September into late fall.

Field Characteristics
Distinctive, blocky bark. Pith of twigs is solid or sometimes chambered. Fruits astringent/bitter, but edible usually after first frost.

Habitat/Range
Forested wetlands, wet fields, dry woodlands. Most common in the Piedmont; less common in the Mountains and Coastal Plain.

Similar Species
Nyssa sylvatica (blackgum) has similar leaves, but *Diospyros virginiana* (persimmon) contains one linear bundle scar when leaf is pulled away from stem, whereas *N. sylvatica* has 3 short bundle scars.

35

Fraxinus caroliniana — Carolina Ash

Often growing as a small, multi-trunk tree

Carolina ash leaves rounder than green ash leaves

Opposite, pinnately compound leaves

Seed samaras wide and flat

Fraxinus seedling leaves often are not compound

TREES

Oleaceae - Olive Family

Fraxinus caroliniana
Carolina Ash

National Wetland Plant List
Mtn/Pdmt: **OBL**   CP: **OBL**

Coefficient of Conservatism
Mtn: **n/a**   Pdmt: **7**   CP: **7**

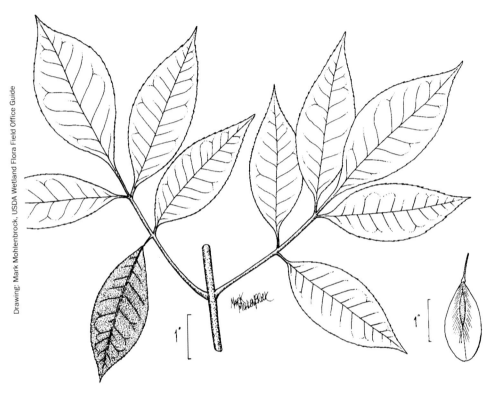

Habit
Small deciduous tree to 8 m, with smooth bark, often with several trunks.

Leaves
Opposite, pinnately compound: 5-9 oval or lance-shaped leaflets with entire or irregularly toothed margins. Seedling leaves are often not compound.

Flowers/Fruit
Small flowers in dense clusters, appearing at or before leaf-out; male and female growing on separate trees. Fruit is a flat widened oval shaped samara. Flowers mainly in May; fruits July to October.

Field Characteristics
Samaras oval shaped and flat or multi-winged, often bright violet.

Habitat/Range
Grows only in the deeper swamps along river bottoms in the Coastal Plain. Common in the Coastal Plain but rare in the Piedmont. Stands are rapidly declining because of the emerald ash borer, which strongly attacks this species, causing much mortality.

Similar Species
Note opposite branches, twigs more slender than hickories which have similar compound leaves, but have alternate branches. *F. caroliniana* (Carolina ash) can be difficult to distinguish from *F. pennsylvanica* (green ash) without fruit, especially in the Coastal Plain where both species are common.

Fraxinus pennsylvanica — Green Ash

Medium to large tree; plated bark

Lance-shaped leaflets; light green samaras

Opposite, pinnately compound leaves

Elongated samaras, late spring and summer

Fraxinus seedling leaves often are not compound

TREES

Oleaceae - Olive Family

Fraxinus pennsylvanica
Green Ash

National Wetland Plant List
Mtn/Pdmt: **FACW**   CP: **FACW**

Coefficient of Conservatism
Mtn: **n/a**   Pdmt: **5**   CP: **5**

Drawing: Karen Kendig

Habit
Medium to large deciduous tree up to 24 m, generally with a single trunk.

Leaves
Opposite, pinnately compound leaves, 15 to 23 cm long. Contain 5-9 oval or lance-shaped toothed leaflets, light green beneath. Seedling leaves are often not compound.

Flowers/Fruit
Flowers inconspicuous with male and female flowers on separate trees. Fruit is a long, very narrow, light green samara. Flowers April/May; fruits August to October.

Field Characteristics
Samaras long and very narrow. Leaves and twigs smooth.

Habitat/Range
Low areas, natural levees, along brownwater rivers and in bottomlands and swamps. This has been the most widely distributed of the ashes, but the emerald ash borer strongly attacks this species, causing much mortality.

Similar Species
F. pennsylvanica (green ash) can be difficult to distinguish from *F. caroliniana* (Carolina ash) without fruit, especially in the Coastal Plain where both species are common. Note opposite branches, twigs more slender than *Carya* spp. (hickories), which have similar compound leaves, but different seeds and alternate branches.

Gordonia lasianthus — Loblolly Bay

Medium evergreen tree

Smooth, reddish, linearly divided bark

Beautiful, showy flowers

Evenly serrated leaf margins; prominent midvein

Dry, splitting fruits

TREES

Theaceae - Tea Family

Gordonia lasianthus
Loblolly Bay

National Wetland Plant List
Mtn/Pdmt: **n/a**   CP: **FACW**

Coefficient of Conservatism
Mtn: **n/a**   Pdmt: **n/a**   CP: **8**

Drawing: Karen Kendig

Habit
Medium evergreen tree up to 20 m. Crown of young tree is narrow and conical, becoming rounded when mature.

Leaves
Alternate, elliptical, dark green, shiny, leathery leaves, 16 cm long and 5 cm wide. Leaf margins are serrated with small, blunt serrations.

Flowers/Fruit
Beautiful 5-petaled white flower with silky fringed stamens in the center. Fruit is a capsule which splits into 5 parts as it releases seeds. Blooms later than many other trees (July to September); fruits September/October.

Field Characteristics
Reddish, smooth bark is distinctive, as well as large showy flowers, when present.

Habitat/Range
Swamps, bay forests, and pocosins in the Coastal Plain, especially extensive pocosins and large Carolina bays.

Similar Species
Similar to other evergreen trees and shrubs in bay forests and pocosins, but leathery leaves of *Gordonia lasianthus* (loblolly bay) are serrated and have reddish petioles.

Ilex opaca — American Holly

Understory tree in moist woods

Smooth light gray bark, often lichen covered

Stamens only present on male trees' flowers

Green fruits on female trees mature to bright red

Dark green leaves have a dull sheen, unlike shiny cultivars

TREES

Aquifoliaceae - Holly Family

Ilex opaca
American Holly

National Wetland Plant List
Mtn/Pdmt: **FACU**   CP: **FAC**

Coefficient of Conservatism
Mtn: **6**   Pdmt: **5**   CP: **5**

Toxic Plant

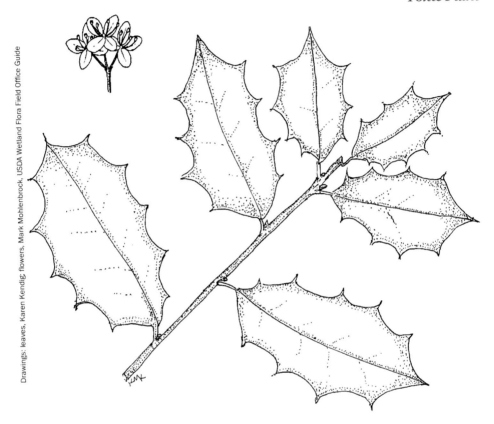

Drawings: leaves, Karen Kendig; flowers, Mark Mohlenbrock, USDA Wetland Flora Field Office Guide

Habit
Small, evergreen, understory tree, usually 5 to 10 m.

Leaves
Leathery, dark green, broadly elliptic leaves with wavy edges and scattered spines along margins.

Flowers/Fruit
Small white 4-petaled flowers appear in late May; trees have either male (with stamens) or female flowers (without stamens). Blooms April to June; fruits September/October. Small red fruits on female trees last through winter.

Field Characteristics
Gray bark smooth and often blotched with various lichen species. This is the only *Ilex* species in North Carolina that grows to a medium sized tree. All hollies have small black stipules at leaf bases. Fruit is toxic.

Habitat/Range
Wide variety of forests, from dry to wetland, but it grows best in moist soil. Less numerous at high elevations in the mountains, but common throughout the state.

Liquidambar styraciflua — Sweetgum

Bark on older tree

Corky, bumpy bark on younger tree

Last year's gum ball and this year's flower

Corky growths on twigs

Some variation in teeth/lobes of leaves

TREES

Hamamelidaceae - Witch Hazel Family

National Wetland Plant List
Mtn/Pdmt: **FAC**   CP: **FAC**

Coefficient of Conservatism
Mtn: **4**   Pdmt: **3**   CP: **3**

Liquidambar styraciflua
Sweetgum

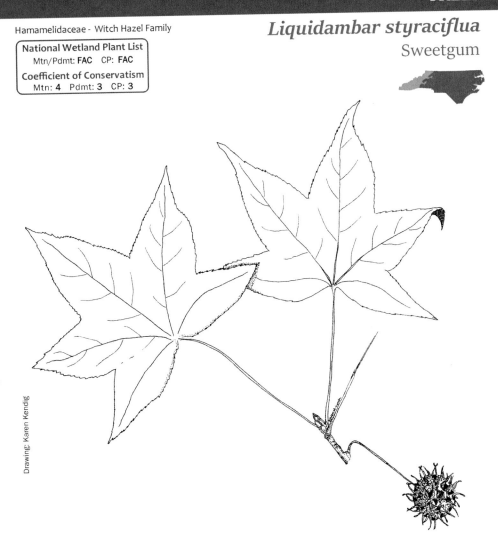

Drawing: Karen Kendig

Habit
Medium to large deciduous tree, to 20 m. Corky growth frequently appears on branches, in any plane.

Leaves
Alternate, palmately lobed (star-shaped) serrated leaves, measuring about 15 cm long by 11 cm wide.

Flowers/Fruit
Fruiting clusters are spherical, woody, spiny, "gum balls", 2-3 cm wide. The gum balls hang like ornaments and persist during winter. Blooms April/May; fruits August/September.

Field Characteristics
Distinctive, ridged corky growth often appears on twigs, but on any side. Star-shaped leaves and prickly gum balls are key features of this nearly unmistakable tree.

Habitat/Range
Swampy woodlands, moist uplands, and old fields. Common throughout NC in a wide range of soil conditions although absent at higher elevations.

Similar Species
In winter, branches may be confused with *Ulmus alata* (winged elm) which has corky growths only along one plane, from two sides of branches.

Liriodendron tulipifera — Tuliptree

Grows very straight and tall

Bark on younger tree

Uniquely shaped leaves; striated bark on older trees

Seeds/samaras spirally arranged on axis

Yellow and orange tulip-like flower

TREES

Magnoliaceae - Magnolia Family

Liriodendron tulipifera
Tuliptree

National Wetland Plant List
Mtn/Pdmt: **FACU**   CP: **FACU**

Coefficient of Conservatism
Mtn: **4**   Pdmt: **4**   CP: **4**

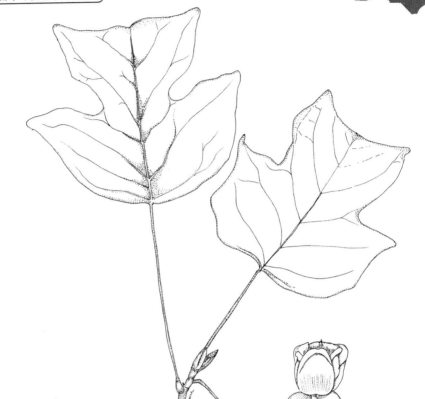

Drawing: Karen Kendig

Habit
Large deciduous tree to 50 m, with towering straight trunk and pointed crown.

Leaves
Alternate, simple, uniquely shaped 4-lobed (or 6-lobed) leaves on long petioles.

Flowers/Fruit
Flower with 9 petals and tulip-shaped. Flowers yellowish-green with a splash of orange. Fruiting cones (aggregate of samaras) persist through winter.

Field Characteristics
As is characteristic of the magnolia family, the stipular scars completely encircle twigs. Tall straight trunk and unique leaf shape are key features for identifying.

Habitat/Range
Low woods, stream sides, headwater seeps, rich moist uplands throughout NC.

Magnolia virginiana — Sweetbay

Semi-evergreen tree to 20 m; silver and green leaves

Rounded, large, fragrant flowers

When mature, fruit splits open to reveal bright red seeds

Leaves with smooth margins; undersides white-silver

Smooth gray bark

TREES

Magnoliaceae - Magnolia Family

National Wetland Plant List
Mtn/Pdmt: **FACW**   CP: **FACW**

Coefficient of Conservatism
Mtn: **n/a**   Pdmt: **7**   CP: **6**

Magnolia virginiana
Sweetbay

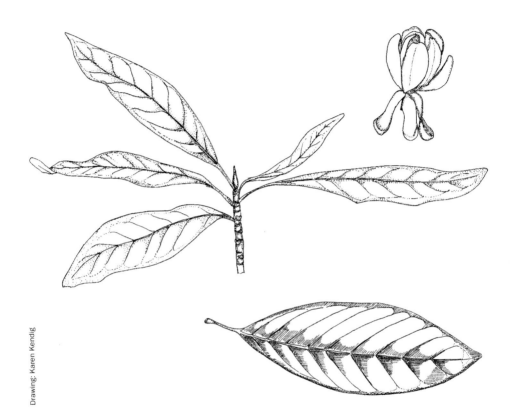

Drawing: Karen Kendig

Habit
Evergreen or semi-evergreen tree or shrub to 20 m. In the northern portion of its range, sweetbay is deciduous.

Leaves
Alternate, entire, long and elliptical or oblong, tapering at the base of the leaf. Leathery leaves 6-15 cm long and 2-6 cm wide with pale undersides.

Flowers/Fruit
Typical "magnolia" flowers with 9-12 white petals; fragrant. Fruit is a dark red cone, 5 cm long. Blooms April to July; fruits July to October.

Field Characteristics
Note stipular scar encircling twig, characteristic of members of the Magnolia family. Long terminal bud is distinctive.

Habitat/Range
Wet flatwoods, swamps, bay forests and savannas in the Coastal Plain. Occasionally found in the Piedmont in moist sandy areas.

Similar Species
Magnolia virginiana (sweetbay) is similar to *Persea palustris* (swamp bay), but *M. virginiana* leaves have white undersides and are not as strongly aromatic.

Nyssa aquatica — Water Tupelo

Tall, straight tree

Flowers on long stems

Green to brownish-purple fruits hang on long stems

Leaves pubescent beneath, sometimes toothed

Buttressed trunks; gray, vertically plated bark; usually found in or near flowing water

TREES

Nyssaceae - Sourgum Family

National Wetland Plant List
Mtn/Pdmt: **OBL**   CP: **OBL**

Coefficient of Conservatism
Mtn: **n/a**   Pdmt: **n/a**   CP: **7**

Nyssa aquatica
Water Tupelo

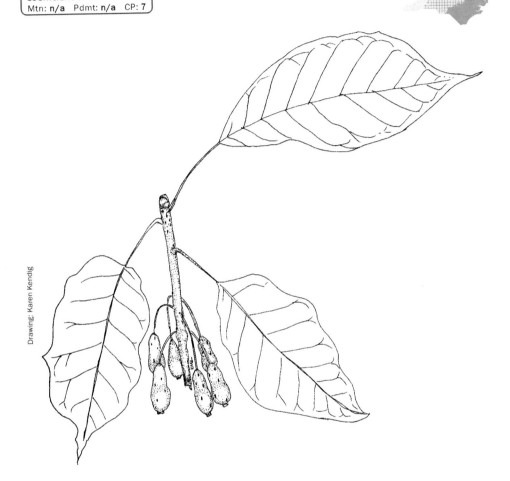

Drawing: Karen Kendig

Habit
Medium to large deciduous tree, to 30 m. Straight trunk typically buttressed when found in regularly flooded areas.

Leaves
Large, ovate or elliptic, alternate and entire or occasionally coarsely toothed. Leaves pubescent, 25 cm long and 15 cm wide, and petioles usually at least 2.5 cm long.

Flowers/Fruit
Elongated, blue-black, 2.5 cm long fruits occur on slender, long (8 cm) drooping stalks. Flowers April/May; fruits September/October.

Field Characteristics
Large leaves for a swamp tree, usually with a few "teeth". Thick twigs and branches with a diaphragmed pith, as in all *Nyssa* species.

Habitat/Range
Floodplain swamps with (at least slowly) flowing water in the Coastal Plain, especially along the Roanoke River, Lumber River, and Waccamaw River. Not often found along creeks or small rivers.

Similar Species
This species is found more often in flowing water than *Nyssa biflora* (swamp tupelo), which has smaller untoothed leaves.

Nyssa biflora — Swamp Tupelo

Leaves widest past middle; shiny above, light green below

Flattened bark has thin, vertical furrows

Usually found in stagnant (not flowing) water

Small fruits singly or in pairs on long stalks

Leaves appear clustered at branch tips

TREES

Nyssaceae - Sourgum Family

National Wetland Plant List
Mtn/Pdmt: **FACW** CP: **OBL**

Coefficient of Conservatism
Mtn: **n/a** Pdmt: **7** CP: **7**

Nyssa biflora
Swamp Tupelo

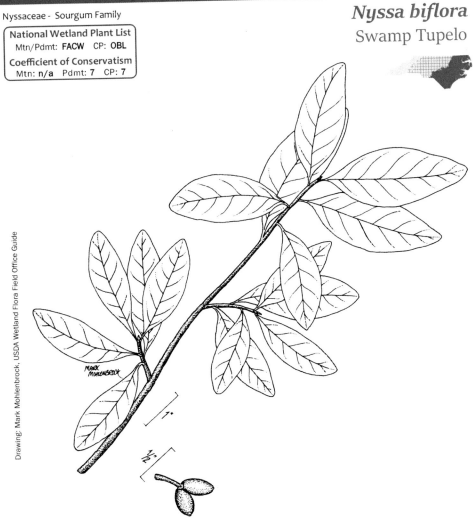

Drawing: Mark Mohlenbrock, USDA Wetland Flora Field Office Guide

Habit
Medium to large deciduous tree, with a buttressed trunk, to 20 m.

Leaves
Alternate, elliptic, shiny, untoothed leaves up to 15 cm long, with a somewhat rounded tip. Leaves appear clustered at branch tips. Individual leaves will sporadically turn red as early as mid-summer, along with *Nyssa sylvatica* (blackgum).

Flowers/Fruit
Male and female flowers occur on separate trees. Fruit is a small, dark blue-black drupe, 1-1.5 cm wide, born singly or in pairs on long stalks. Flowers April to June; fruits August to October.

Field Characteristics
Three bundle scars revealed when leaf is pulled away from stem. Leaves turn a dark red in fall, some earlier. Flattened bark has thin, vertical furrows. Thick twigs and branches with a diaphragmed pith, as in all *Nyssa* species.

Habitat/Range
Usually found in standing waters in the Coastal Plain, mainly in poorly drained areas. Much more common and widespread than *N. aquatica* (water tupelo).

Similar Species
Nyssa biflora (swamp tupelo) leaves are untoothed and smaller than the rectangular leaves of *N. aquatica* (water tupelo). *N. biflora* has thicker, narrower obovate leaves with more rounded tips than *N. sylvatica* (blackgum).

Nyssa sylvatica — Blackgum

Leaves have pointed tips

Bark on older trees divided into rectangular sections

Leaves may be toothed

Leaves pubescent hairy beneath

Purple-black fruits on long stalks

TREES

Nyssaceae - Sourgum Family

Nyssa sylvatica
Blackgum

National Wetland Plant List
Mtn/Pdmt: **FAC**   CP: **FAC**

Coefficient of Conservatism
Mtn: **6**   Pdmt: **6**   CP: **6**

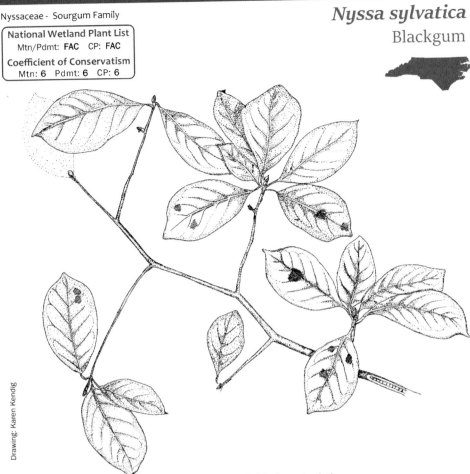

Drawing: Karen Kendig

Habit
Medium to large deciduous tree, to 25 m.

Leaves
Alternate, elliptic leaves, somewhat shiny above; margins usually entire, sometimes toothed on young trees, pubescent beneath. Leaves appear clustered at branch tips. Individual leaves will sporadically turn red as early as mid-summer, along with *Nyssa biflora* (swamp tupelo).

Flowers/Fruit
Separate trees produce male and female flowers. Fruits are egg-shaped, dark blue-black drupes, 1-1.5 cm wide, on long stalk. Flowers April to June; fruits August to October.

Field Characteristics
Three bundle scars revealed when leaf is pulled away from stem. Leaves turn brilliant red-orange in fall, some earlier. Bark on older trees divided into rectangular sections. Thick twigs and branches with a diaphragmed pith, as in all *Nyssa* species.

Habitat/Range
Found in dry uplands and seldomly flooded wetlands statewide, except absent from the northeast part of the state.

Similar Species
Leaves can be confused with *Diospyros virginiana* (common persimmon), but *Nyssa sylvatica* (blackgum) has shinier leaves with scattered teeth. *N. sylvatica* also has wider, thinner leaves with sharper tips than *N. biflora* (swamp tupelo) and can be found outside wetlands. Leaves are a similar shape to *Lindera benzoin* (northern spicebush), which has lemony fragrant, waxy-coated leaves and small red fruits.

Persea palustris — Swamp Bay

Leathery, shiny leaves

Leaves often have galls on margins

Flowers and fruits born on very long stalks

Small, pale flowers

Fine, reddish hairs on new twigs and undersides of leaves

TREES

Lauraceae - Laurel Family

Persea palustris
Swamp Bay

National Wetland Plant List
Mtn/Pdmt: **FACW** CP: **FACW**

Coefficient of Conservatism
Mtn: **n/a** Pdmt: **7** CP: **7**

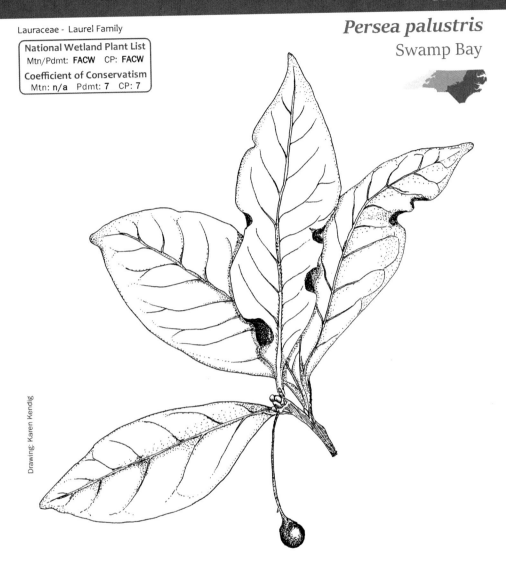

Drawing: Karen Kendig

Habit
Medium sized evergreen tree or shrub.

Leaves
Alternate, entire, dark green leaves with prominent midvein above, white undersides, and spreading reddish hairs along veins. Leaves typically have galls on edges.

Flowers/Fruit
Small inconspicuous, white flowers, later form spherical green berries which turn blue-black upon maturity. Blooms May/June; fruits September/October.

Field Characteristics
Twigs densely pubescent. Thick leaves strongly aromatic when crushed.

Habitat/Range
Swamps, pocosins, bay forests, and moist sandy areas, mainly in the Coastal Plain and lower Piedmont. Usually in wet peat soils, but also in drier maritime forests.

Similar Species
Persea palustris (swamp bay) has pubescent twigs, whereas the less common *Persea borbonia* (upland redbay) has smooth twigs, leaves without hairy veins, and is found in drier areas.

Pinus palustris — Longleaf Pine

Straight trees; branch tip needles in cylindrical tufts

Very long needles distinguish this pine

Bottlebrush stage

Very large cones

Grass stage

TREES

Pinaceae - Pine Family

National Wetland Plant List
Mtn/Pdmt: **FAC** CP: **FAC**

Coefficient of Conservatism
Mtn: **n/a** Pdmt: **8** CP: **7**

Pinus palustris
Longleaf Pine

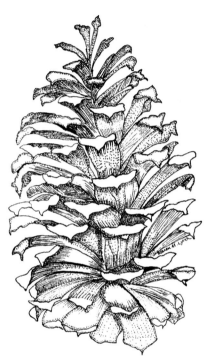

Drawing: Karen Kendig

Habit
Elegant tall evergreen tree. Seedlings resemble clumps of grass.

Leaves
Long needles, 25-40 cm in bundles of 3. Needles arranged in dense spherical tufts near the ends of thick twigs.

Flowers/Fruit
Cones large, 20-45 cm long and brown and prickly. Pollen released March/April; cones mature September/October.

Field Characteristics
Long, graceful needles and large cones distinguish this from other pines. New growth buds are distinctively large and silvery white, a highly visible trait in winter. This is a long-lived tree that thrives when periodic fires burn its habitat.

Habitat/Range
Sandy soil in the Coastal Plain and adjacent Piedmont; most numerous in dry sands, but ranges to wet soils (savannas).

Similar Species
Pinus taeda (loblolly pine) has thin twigs and clusters of needles at branch tips in cylindrical, not spherical tufts.

Pinus serotina — Pond Pine

Blocky, brown bark

Usually somewhat crooked or gnarled habit

Mainly in the Coastal Plain; along lakes, in pocosins

Distinctive flat-ended cone

Needles usually in bundles of 3, growing in all directions

TREES

Pinaceae - Pine Family

National Wetland Plant List
E. Mtn: **OBL**   CP: **FACW**

Coefficient of Conservatism
Mtn: **n/a**   Pdmt: **9**   CP: **8**

Pinus serotina
Pond Pine

Drawing: Karen Kendig

Habit
Medium sized evergreen tree, with a gnarled appearance.

Leaves
Needles 10-20 cm long, usually in bundles of 3, growing in all directions.

Flowers/Fruit
Top-shaped closed cones remain on the tree for several years. Pollen released in April; cones mature in August. Fire forces cones to open and release seeds.

Field Characteristics
Tufts of twigs and needles often found growing from trunks, especially after fires. Of the pines, pond pine is the most tolerant of wet conditions.

Habitat/Range
Found in acidic, peat-based soils of pocosins and in wet flats along lake edges, mainly in the Coastal Plain.

Similar Species
Easily confused with *Pinus taeda* (loblolly pine), but best distinguished by rounded cones almost as wide as long, as well as general crooked appearance.

Pinus taeda — Loblolly Pine

Brown bark with large rectangular plates

Spherical bunches of needles at branch tips

Prickly female cones 10 cm long

Needles in bundles of three; thin twigs

Yellow male cones release large amounts of pollen in spring

TREES

Pinaceae - Pine Family

National Wetland Plant List
Mtn/Pdmt: **FAC**   CP: **FAC**

Coefficient of Conservatism
Mtn: **n/a**   Pdmt: **2**   CP: **2**

Pinus taeda
Loblolly Pine

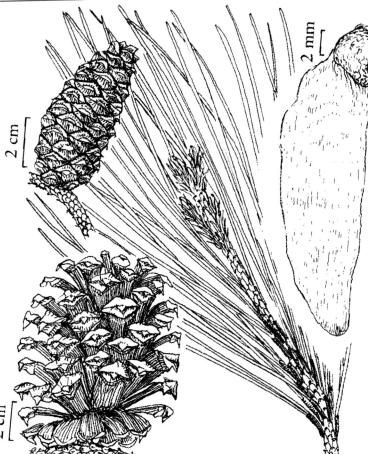

Drawing: Courtesy of the Flora of North America Association, K. Smith

Habit
Medium to large evergreen tree.

Leaves
Needles in bundles of 3 and 15-20 cm long.

Flowers/Fruit
Prickly, brown female cones remain on the tree for 3 years.; about 10 cm long. Less conspicuous male pollen bearing cones found at branch ends. Pollen released in profusion March/April; cones mature October/November.

Field Characteristics
Often called "old field pine" because of its tendency to invade abandoned fields. Cones approximately twice as long as wide. Branch tips have spherical bunches of needles.

Habitat/Range
Wet flats, old fields, in the Coastal Plain and Piedmont of NC; very tolerant of range in moisture levels, though scarce in deep sands.

Similar Species
Pinus palustris (longleaf pine) has thick twigs and clusters of needles at branch tips in cylindrical tufts, not spherical like *P. taeda* (loblolly pine). Also confused with *Pinus serotina* (pond pine), which has flat-ended cones.

Platanus occidentalis — American Sycamore

Bark of older tree

Bark of younger tree

Typical leaf shape

Seed balls persist through winter

Large leaves, to 25 cm long and wide

TREES

Platanaceae - Sycamore Family

National Wetland Plant List
Mtn/Pdmt: **FACW**   CP: **FACW**

Coefficient of Conservatism
Mtn: **4**   Pdmt: **5**   CP: **5**

Platanus occidentalis
American Sycamore

Drawing: Karen Kendig

Habit
Large deciduous tree to 35 m. Bark on older trees white and plate-like, often flaking off.

Leaves
Alternate, toothed leaves up to 25 cm long and wide. Leaf petiole has swollen base and crown-like stipules present at the point of attachment.

Flowers/Fruit
The brown spherical fruiting 'head' is 2-3 cm in diameter, contains many seeds and persists through winter. Seeds released in spring. Flowers April to June; fruits September to November.

Field Characteristics
Bark on older trees forms a beautiful mottled patchwork of white, gray, green and yellow. Upper bark of older trees is strikingly white.

Habitat/Range
Flood plains, low moist woods, edges of lakes and streams throughout NC, less commonly in the eastern Coastal Plain.

Similar Species
Acer spp. (maples) have leaves of similar shape, but generally much smaller, with bark that does not flake off.

Quercus laurifolia — Laurel Oak

Relatively smooth bark; medium to large tree

Semi-evergreen leaves

Tufts of hair along midrib on leaf undersides

Small, rounded acorns take 18 months to mature

Leaves generally widest at or past middle; glossy surface

TREES

Fagaceae - Beech Family

Quercus laurifolia
Laurel Oak

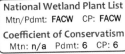

National Wetland Plant List
Mtn/Pdmt: **FACW**   CP: **FACW**

Coefficient of Conservatism
Mtn: **n/a**   Pdmt: **6**   CP: **6**

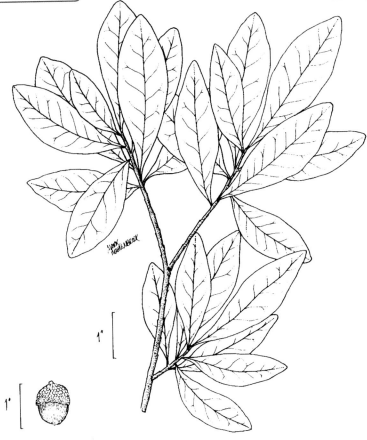

Drawing: Mark Mohlenbrock, USDA Wetland Flora Field Office Guide

Habit
Medium to large (to 30 m) semi-evergreen tree with leaves gradually dropping in late fall and winter.

Leaves
Alternate, narrowly ovate or oblanceolate leaves, widest past the middle, often persisting through winter. Thick leaves have lustrous surface, 3-9 cm long and 2-3 cm wide, entire margins and blunt tips with a short bristle.

Flowers/Fruit
Acorns faintly striped with shallow, bowl-like cups enclosing one third of the acorn. Blooms March/April; acorns mature September to November of following year.

Field Characteristics
Leaves narrower and less leathery than *Quercus virginiana* (live oak) and wider than *Q. phellos* (willow oak). Leaves of seedlings may be 3-lobed and differ greatly from leaves of a mature tree.

Habitat/Range
Floodplain forests, stream banks, black and brownwater swamps, mostly in the Coastal Plain.

Similar Species
Quercus phellos (willow oak) has similar leaves, but they are narrower and have bristles on the tips. *Q. hemisphaerica* (Darlington oak) is a very similar tree, but is found in dry sandy Coastal Plain soils.

Quercus lyrata — Overcup Oak

Straight, medium to large deciduous tree

Narrowly plated bark

Leaves quite narrow in middle

Large, encased acorns; mature in one summer

Laura Clark, inaturalist

Leaves 12-23 cm long; up to 12 cm wide

TREES

Fagaceae - Beech Family

Quercus lyrata
Overcup Oak

National Wetland Plant List
Mtn/Pdmt: **OBL** CP: **OBL**

Coefficient of Conservatism
Mtn: **n/a** Pdmt: **7** CP: **7**

Drawing: Karen Kendig

Habit
Medium to large deciduous tree up to 30 m tall.

Leaves
Alternate, usually 7-lobed and obovate in general outline, although variable. Leaves pale on undersides; 12-23 cm long and up to 12 cm wide.

Flowers/Fruit
Acorns globose or slightly flattened with the nut almost completely covered by a ragged scaly cup, earning the name, "overcup." Flowers March/April; acorns mature September/October of the same year.

Field Characteristics
Note lighter undersides of 7-lobed leaves and distinctive acorns when present. Leaves quite narrow in middle.

Habitat/Range
Bottomlands, swamps, floodplains, and ephemeral wetlands in the Coastal Plain and Piedmont.

Quercus michauxii — Swamp Chestnut Oak

Large deciduous tree at maturity

Flattened-scaly bark

Twigs and leaves usually pubescent

Large 3 cm long acorns; stems 1 cm or less

Large obovate leaves with crenate or shallow-lobed margins, typically about 18 cm long and 10 cm wide

TREES

Fagaceae - Beech Family

National Wetland Plant List
Mtn/Pdmt: **FACW**   CP: **FACW**
Coefficient of Conservatism
Mtn: **n/a**   Pdmt: **7**   CP: **7**

Quercus michauxii
Swamp Chestnut Oak

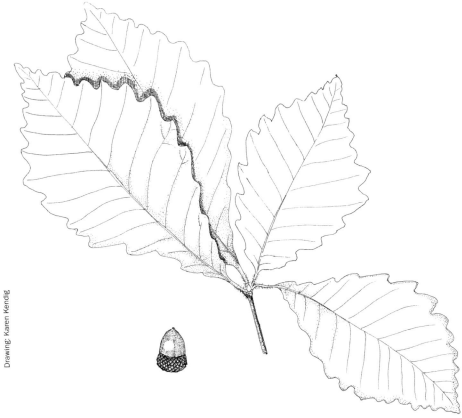

Drawing: Karen Kendig

Habit
Large deciduous tree, to 25 m, with thick branches projecting at sharp angles to form a round-topped crown.

Leaves
Alternate, obovate (in general outline) with crenate or shallow-lobed margins. Top of leaves dark green but not shiny, and undersides gray to rust-colored and sparsely pubescent. Leaf size variable; typically about 18 cm long and 10 cm wide.

Flowers/Fruit
Acorn relatively large (about 3 cm long) with cup enclosing a third of the nut; acorn stem rarely more than 1 cm. Flowers April/May; acorns mature September/October of the same year.

Field Characteristics
Distinctive thin, crenate leaves. This oak has light grey, flattened-scaly bark and leaves are usually pubescent.

Habitat/Range
Brownwater floodplain forests, ephemeral wetlands, seldom flooded wet woods in the Coastal Plain and lower Piedmont. Usually absent from blackwater floodplains.

Similar Species
Similar to *Quercus montana* (chestnut oak), which grows in rocky uplands and has dark, deeply furrowed bark. Similar to the rare *Quercus bicolor* (swamp white oak), which has thick, shiny, slightly more deeply lobed leaves with pale green or greenish-white undersides. Acorns of this species have long stems.

Quercus nigra — Water Oak

Deciduous tree of bottomlands and floodplains

Gray, fissured bark

Triangular or diamond-shaped top half of leaves

Small acorns take 18 months to develop

Flowers in April, Fruits Sept/Nov of following year

TREES

Fagaceae - Beech Family

> **National Wetland Plant List**
> Mtn/Pdmt: **FAC**   CP: **FAC**
>
> **Coefficient of Conservatism**
> Mtn: **4**   Pdmt: **3**   CP: **3**

Quercus nigra
Water Oak

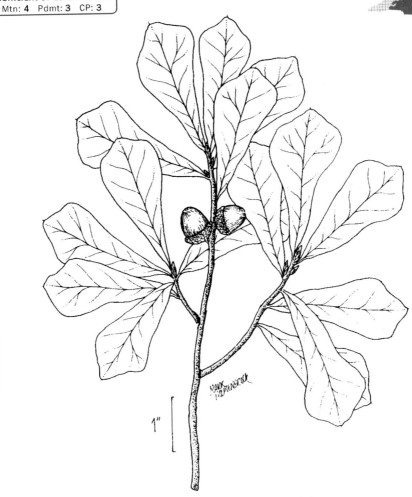

Drawing: Mark Mohlenbrock, USDA Wetland Flora Field Office Guide

Habit
Medium sized tree, up to 25 m. Although deciduous, leaves are slow to fall, sometimes remaining into winter.

Leaves
Alternate, simple, club-shaped or 3-lobed but highly variable. Leaves widest toward the tip, obovate in general outline and about 10 cm long.

Flowers/Fruit
Dark, oval acorns are about 1 cm wide and only a third covered by the saucer-like cup. Inside, cup is shiny-pubescent. Flowers in April; acorns mature September to November of following year.

Field Characteristics
Leaves commonly pear shaped but vary considerably. May be tardily deciduous.

Habitat/Range
Bottomlands, brownwater floodplains, and occasionally blackwater floodplains, moist soils, and wet flats in the Coastal Plain and Piedmont. Occasionally found in the easternmost mountains.

Similar Species
Quercus phellos (willow oak) has similar leaves that are narrower, with bristles on the tips.

Quercus pagoda — Cherrybark Oak

Tall, straight tree

Bark with linear divisions and buttressed trunk

Soft, pale, tawny brown hairs on leaf undersides

Small acorns take 18 months to mature
Patrick Meagher, iNaturalist

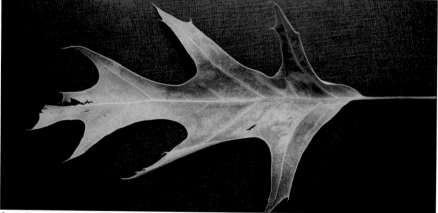
Some leaves more deeply lobed than others

TREES

Fagaceae - Beech Family

Quercus pagoda
Cherrybark Oak

National Wetland Plant List
Mtn/Pdmt: **FACW**   CP: **FAC**

Coefficient of Conservatism
Mtn: **n/a**   Pdmt: **7**   CP: **7**

Drawing: Karen Kendig

Habit
Large straight deciduous tree, to 40 m.

Leaves
Alternate, obovate to ovate in general outline, 10-20 cm long and 8-14 cm wide, typically with 5 lobes, sometimes more. End lobe usually toothed and spaces between lobes generally v-shaped (not rounded). Backs of leaves whitish and densely pubescent.

Flowers/Fruit
Flowers appear in spring, when leaves first emerge. Small, rounded acorns short-stalked and 1 cm long in cup-like saucers. Flowers April/May; acorns mature September to November of following year.

Field Characteristics
Leaves with sharply angled spaces between lobes and v-shaped bases. Leaf undersides densely white pubescent.

Habitat/Range
Low grounds including brownwater floodplains and bottomland woods, occasionally on slopes and bluffs. Coastal Plain and lower Piedmont.

Similar Species
Quercus falcata (southern red oak) has more rounded (bell-shaped) leaf base and less pale, more tan-colored pubescence on leaves.

Quercus phellos — Willow Oak

Bristles on tips of linear leaves

Gray, furrowed bark

Clusters of short white hairs along midrib beneath

Small acorns take 18 months to mature

Leaves are semi-evergreen

TREES

Fagaceae - Beech Family

Quercus phellos
Willow Oak

National Wetland Plant List
Mtn/Pdmt: **FAC**   CP: **FACW**

Coefficient of Conservatism
Mtn: **n/a**   Pdmt: **5**   CP: **5**

Drawing: Karen Kendig

Habit
Medium to large deciduous tree, up to 30 m. May be semi-evergreen in southernmost localities.

Leaves
Alternate, entire, linear or linear-lanceolate with short bristle tips. Leaves typically 9 cm long and less than 2 cm wide, though sometimes larger.

Flowers/Fruit
Yellowish or greenish-brown acorn, about 1 cm long with only the base of nut enclosed by cup. Flowers March to May; acorns mature from September to November of following year.

Field Characteristics
"Willow-like" leaves taper at both ends with bristles on tips. Leaves turn brown by mid to late fall.

Habitat/Range
Brownwater floodplains, forested wetlands, floodplain pools, and depression wetlands in the Coastal Plain and Piedmont. Can occur in blackwater floodplains. Widely planted in landscaping.

Similar Species
Quercus laurifolia (laurel oak) has more diamond shaped leaves and generally without bristles on the tips. *Q. nigra* (water oak) leaves are much wider near the tips.

Salix nigra — Black Willow

Long, linear leaves, greenish beneath; female catkins

Rough, scaly gray bark

Usually grows as a few-trunked, small tree

Catkins emerge in spring

Female catkins produce seeds borne on cottony fibers

TREES

Salicaceae - Willow Family

Salix nigra
Black Willow

National Wetland Plant List
Mtn/Pdmt: **OBL**   CP: **OBL**

Coefficient of Conservatism
Mtn: **4**   Pdmt: **3**   CP: **3**

Drawing: Karen Kendig

Habit
Deciduous small tree, to 15 m high.

Leaves
Alternate, finely toothed, narrowly lance-shaped or sickle-shaped leaves, 12 cm long and 1-2 cm wide. Leaves dark green and shiny with stipules sometimes present at base of leaves; greenish beneath.

Flowers/Fruit
Male and female flowers on separate catkins. Fruit a pod-bearing seed attached to a cottony mass for easy seed dispersal. Flowers and fruits March/April.

Field Characteristics
Twigs in winter conspicuously red, yellowish or green. Willows root easily, making them popular for use in stream restoration projects.

Habitat/Range
Along streams, wet woodlands, fresh marshes, swamps and floodplains throughout the state although absent in high mountain elevations.

Similar Species
Salix caroliniana (Coastal Plain willow) has leaves whitish beneath and grows as a multi-stemmed, large shrub; *S. nigra* (black willow) has leaves that are green beneath and often grows as a small tree with a few trunks.

Taxodium ascendens — Pond Cypress

Swollen trunk base; light gray bark soft and shredding

Appressed needles on older trees

Round crinkled cones mature from green to brown

Needles flat in young pond cypress trees

Branchlets often oriented upwards in older trees

TREES

Cupressaceae - Cypress Family

Taxodium ascendens
Pond Cypress

National Wetland Plant List
Mtn/Pdmt: **OBL**   CP: **OBL**

Coefficient of Conservatism
Mtn: **0**   Pdmt: **8**   CP: **8**

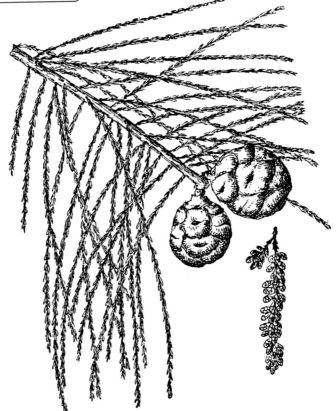

Habit
Medium to large deciduous conifer with wide spreading base, especially when growing in water.

Leaves
Needles on grown trees short and pressed together ("appressed") along upward pointing branchlets.

Flowers/Fruit
Female cones ball-shaped with brown, scale-like markings. Pollen released March/April; fruits in October.

Field Characteristics
Needles short and pressed together. Bark thick, soft, and shredding.

Habitat/Range
Pond cypress grows mainly in the Coastal Plain and Sandhills, in still-water areas such as Carolina bays, pocosins and other wet peaty habitats, shores of natural blackwater lakes, and non-riverine swamps.

Similar Species
Taxodium distichum (bald cypress) has flat needles, compared to appressed needles in *T. ascendens* (pond cypress); however, needles on young trees and new shoots of *T. ascendens* appear more like those of *T. distichum* (bald cypress). Bark soft and shredding, unlike *T. distichum*.

Taxonomic Note
Many references consider the two cypresses to be varieties of the same species, *T. distichum*, with visual differences are attributed to environmental factors

Taxodium distichum — Bald Cypress

Widely buttressed trunk; light gray bark

Bark rough and not shredding

Needles flat in young and older cypress

Catkins appear in early spring

Ball-shaped cones mature from green to brown

TREES

Cupressaceae - Cypress Family

Taxodium distichum
Bald Cypress

National Wetland Plant List
Mtn/Pdmt: **OBL**   CP: **OBL**

Coefficient of Conservatism
Mtn: 0   Pdmt: 6   CP: 6

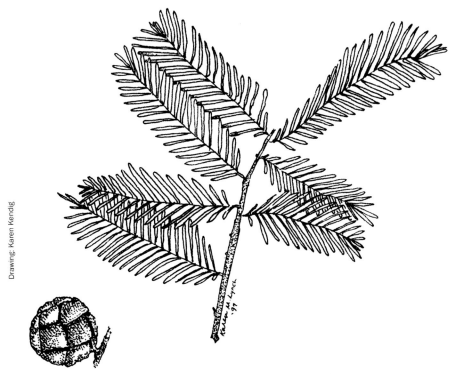

Drawing: Karen Kendig

Habit
Medium to large deciduous conifer with straight trunk, horizontal branches, and wide spreading base. Crowns of young trees are conical, but become "flat-topped" with old age. Heights to 40 m or more.

Leaves
Narrow linear leaves to 2 cm long, occurring in one plane and appearing feather-like on small alternate branches that are not upturned.

Flowers/Fruit
Female cones ball-shaped with brown scale-like markings. Male cones less conspicuous in drooping panicles. Pollen released March/April; fruits in October.

Field Characteristics
Knees and buttressed trunks characteristic of cypress. Note flat-topped shape of older trees. Bark thin, rough, and not shredding.

Habitat/Range
Blackwater and brownwater rivers, swamps, forested wetlands, edges of ponds, mainly in the Coastal Plain. Although it cannot germinate in water, cypress will thrive in open water once established.

Similar Species
Taxodium ascendens (pond cypress) needles are smaller, shorter, and appressed on upturned branchlets; however, needles on young trees and new shoots of *T. ascendens* appear more like those of *T. distichum*. Needles are superficially similar to *Tsuga canadensis* (eastern hemlock), which is found in the mountains but never in the Coastal Plain.

Taxonomic Note
Many references consider the two cypresses to be varieties of the same species, *T. distichum*, with visual differences attributed to environmental factors.

Ulmus americana — American Elm

Rough, scaly bark

Base often spreading or buttressed

Asymmetrical leaf base; prominent, straight veins

Zigzag branching pattern; fuzzy samaras

Doubly serrated leaves, rough in one direction above

TREES

Ulmaceae - Elm Family

National Wetland Plant List
Mtn/Pdmt: **FACW** CP: **FAC**

Coefficient of Conservatism
Mtn: **5** Pdmt: **6** CP: **6**

Ulmus americana
American Elm

Drawings: leaves - Karen Kendig; fruit - Mark Mohlenbrock, USDA Wetland Flora Field Office Guide

Habit
Medium to large deciduous tree, to 35 m. Crown broad and spreading, with a characteristic vase-like pattern of branching, unless found in dense forest stands, where tree exhibits a narrow crown.

Leaves
Alternate, doubly serrated, oval leaves with distinctly asymmetrical leaf base (as in all elms). Leaves rough in one direction above, less so beneath. Average size about 8 cm long, 5 cm wide, but highly variable. Bark ridged and scaly.

Flowers/Fruit
Fruit small, flattened clusters of fuzzy, oval samaras, about 1 cm long. Flowers February/March, fruits March/April.

Field Characteristics
Scaly bark and asymmetrical leaf base of American elm are distinctive.

Habitat/Range
Most common in bottomland floodplains adjacent to brownwater streams, rich wet or upland woodlands throughout NC.

Similar Species
Ulmus americana (American elm) leaves are smaller and not as rough as *U. rubra* (slippery elm), but larger and hairier than *U. alata* (winged elm). Upper leaf surface in this elm is smoother than other similar elms. Can be confused with *Betula* spp. (birches); see Common Confusions section p. 402.

SHRUBS

Scientific Name	Common Name	National Wetland Plant List Status E. Mtns & Piedmont/Coastal Plain	Page
Aesculus sylvatica	Painted Buckeye	FAC/FAC	88
Alnus serrulata	Tag Alder	OBL/FACW	90
Aronia arbutifolia	Red Chokeberry	FACW/FACW	92
Asimina triloba	Common Pawpaw	FAC/FAC	94
Baccharis halimifolia	Eastern Baccharis	FACW/FAC	96
Borrichia frutescens	Sea Ox-Eye Daisy	(n/a)/OBL	98
Cephalanthus occidentalis	Common Buttonbush	OBL/OBL	100
Clethra alnifolia	Coastal Sweet-pepperbush	FAC/FACW	102
Cornus amomum	Silky Dogwood	FACW/FACW	104
Cyrilla racemiflora	Swamp Titi	FACW/FACW	106
Decodon verticillatus	Swamp Loosestrife	OBL/OBL	108
Eubotrys racemosus	Swamp Fetterbush	FACW/FACW	110
Euonymus americanus	American Strawberry-bush	FAC/FAC	112
Gaylussacia frondosa	Blue Huckleberry	FAC/FAC	114
Hibiscus moscheutos	Swamp Rose Mallow	OBL/OBL	116
Hypericum hypericoides	St. Andrew's-Cross	FACU/FAC	118
Ilex coriacea	Large Gallberry	FACW/FACW	120
Ilex decidua	Possumhaw	FACW/FACW	122
Ilex glabra	Inkberry	FAC/FACW	124
Ilex verticillata	Common Winterberry	FACW/FACW	126
Itea virginica	Virginia Sweetspire	OBL/FACW	128
Iva frutescens	Marsh Elder	(n/a)/FACW	130
Kosteletzkya virginica	Saltmarsh Mallow	(n/a)/OBL	132
Leucothoe axillaris	Coastal Doghobble	FACW/FACW	134
Ligustrum sinense	Chinese Privet	FACU/FAC	136
Lindera benzoin	Northern Spicebush	FAC/FACW	138
Lyonia ligustrina	Maleberry	FACW/FACW	140
Lyonia lucida	Fetterbush Lyonia	FACW/FACW	142
Morella caroliniensis	Southern Bayberry	FAC/FACW	144
Morella cerifera	Common Wax Myrtle	FAC/FAC	146
Rosa palustris	Swamp Rose	OBL/OBL	148
Rubus pensilvanicus	Sawtooth Blackberry	FAC/FAC	150
Salix caroliniana	Coastal Plain Willow	OBL/OBL	152
Sambucus nigra	Black Elderberry	FAC/FACW	154
Symplocos tinctoria	Common Sweetleaf	FAC/FAC	156
Vaccinium corymbosum	Highbush Blueberry	FACW/FACW	158
Viburnum nudum	Possumhaw Viburnum	OBL/FACW	160
Viburnum recognitum	Southern Arrowwood	FAC/FAC	162

Aesculus sylvatica — Painted Buckeye

Tubular flowers appear late March to early May

Understory shrub of floodplains and rich slopes

Fruits appear July/August and remain after leaves drop

Five-part compound leaves with straight veins

One of the first shrubs to leaf out in spring

SHRUBS

Hippocastanaceae - Soapberry Family

National Wetland Plant List
Mtn/Pdmt: **FAC**   CP: **FAC**

Coefficient of Conservatism
Mtn: **6**   Pdmt: **6**   CP: **6**

Aesculus sylvatica
Painted Buckeye

Highly Toxic Seed

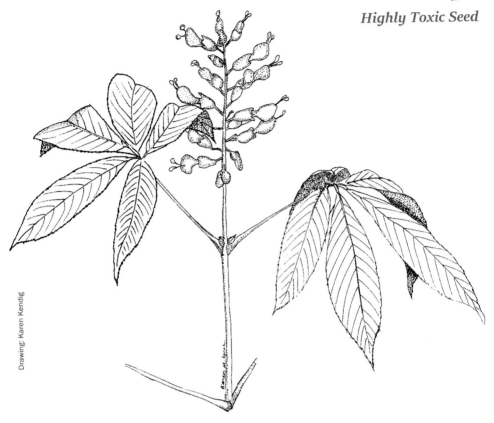

Drawing: Karen Kendig

Habit
Deciduous understory shrub up to 3 m tall, rarely to 10 m.

Leaves
Oppositely arranged, palmately compound leaves with 5-7 leaflets. Leaflets 8-20 cm long and 3-7 cm wide. Leaflets lance-shaped or obovate.

Flowers/Fruit
Tubular flowers (cream, yellow, or pink) on stalk. Buckeye seeds produced inside thick capsule. Flowers late March to early May; fruits July/August.

Field Characteristics
This shrub is usually the first to leaf out in early spring and the first to drop leaves in late summer or early fall. The "buckeye" seed is poisonous to humans if ingested.

Habitat/Range
Rich woods, river banks and floodplains, mainly in the Piedmont.

Similar Species
Aesculus pavia (red buckeye) is a similar uncommon shrub along blackwater streams in the Coastal Plain. *A. flava* (yellow buckeye) is a common, tall tree found in northern hardwood forests and on cool, moist slopes. Can be confused with *Asimina triloba* (common pawpaw); see Common Confusions section, p. 405.

Alnus serrulata — Tag Alder

Large bush or small multi-trunk tree

Finely toothed leaf margins

Mature cone-like fruit develop from female cones

Male catkins

Prominent, straight veins in ovate leaves

SHRUBS

Betulaceae - Birch Family

National Wetland Plant List
Mtn/Pdmt: **OBL**   CP: **FACW**

Coefficient of Conservatism
Mtn: **5**   Pdmt: **5**   CP: **5**

Alnus serrulata
Tag Alder

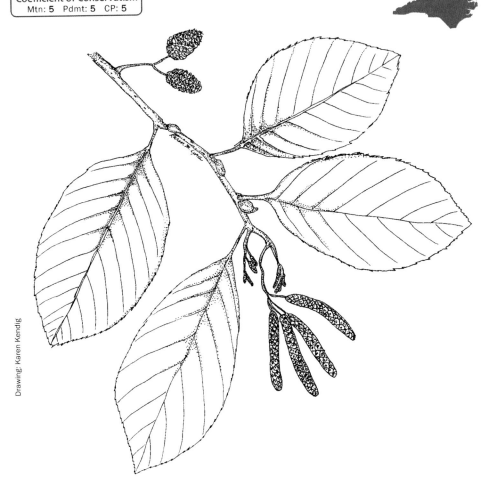

Drawing: Karen Kendig

Habit
Multiple stemmed deciduous shrub up to 5 m in height. Twigs rusty and pubescent, later becoming smooth.

Leaves
Alternate, ovate or obovate shaped with toothed margins. Leaves 6-10 cm long and 2.5-5 cm wide.

Flowers/Fruit
Both male and female flowers occur on shrub on stalks. Female "flowers" or catkins appear as miniature cones. Male catkins initially short and eventually dangling. Blooms February/March; fruits August to October.

Field Characteristics
Female "cones" persist through winter.

Habitat/Range
Streambanks, freshwater marshes, and swamps throughout NC.

Aronia arbutifolia — Red Chokeberry

Medium shrub, 2 to 3 m

Twigs and backs of leaves pubescent

Finely toothed margins around entire leaf

Pinkish-white 5-petaled flowers at branch tips

Immature fruits later mature into red, berry-like pomes

SHRUBS

Rosaceae - Rose Family

National Wetland Plant List
Mtn/Pdmt: **FACW**   CP: **FACW**

Coefficient of Conservatism
Mtn: **7**   Pdmt: **7**   CP: **7**

Aronia arbutifolia
Red Chokeberry

Habit
Low to medium deciduous shrub, 2-3 m in height. Spreads by rhizomes and can form dense colonies.

Leaves
Alternate, simple, elliptic leaves with finely toothed margins. Leaves 4-10 cm long and 2-4 cm wide. Leaves contain minute reddish-brown hairs along midrib vein.

Flowers/Fruit
Clusters of small pinkish-white 5-petaled flowers produced on the ends of branches. Fruit is red, berry-like pome. Blooms late February to May; fruits September to November.

Field Characteristics
Leaves burgundy in fall and red fruits persist in winter.

Habitat/Range
Low woodlands, pine savannas, creek banks, seepage slopes and swamps. Common statewide.

Taxonomic Note
Synonym: *Photinia pyrifolia*

Asimina triloba — Common Pawpaw

Deciduous understory shrub or small tree to 10 m

Fleshy, edible fruits turn yellow with brown seeds

Leaves malodorous when crushed

Red sepals on flowers, appearing before leaves

Oblanceolate leaf shape

SHRUBS

Annonaceae - Custard Apple Family

Asimina triloba
Common Pawpaw

National Wetland Plant List
Mtn/Pdmt: **FAC**   CP: **FAC**

Coefficient of Conservatism
Mtn: **6**   Pdmt: **7**   CP: **7**

Toxic Plant

Habit
Deciduous understory shrub or small tree up to 10 m, often forming colonies.

Leaves
Alternate, entire, oblanceolate leaves, with acuminate tips, about 23 cm long and 8 cm wide. Leaves malodorous when crushed.

Flowers/Fruit
Large (3-4 cm) burgundy flowers with 6 petals and 3 burgundy sepals. Fruits fleshy, edible, yellow and banana-like with large brown seeds, up to 12 cm long. Flowers March to May, before leaves; fruits August to October.

Field Characteristics
Usually an understory shrub. Distinctive acrid odor of crushed leaves helps in recognition, along with flowers and fruits when present.

Habitat/Range
Rich slopes, low woods, bottomlands; within the Coastal Plain, more common in the inner portion, along brownwater rivers.

Similar Species
Leaves can be confused with *Aesculus sylvatica* (painted buckeye); see Common Confusions section, p. 405.

Baccharis halimifolia — Eastern Baccharis

Blooms in fall (male plant - left; female plant - right)

Semi-evergreen woody shrub with ascending branches

Distinctive toothed, obovate or elliptic leaf

Male flowers with shorter bristles

Linear, untoothed leaves near flowers/fruits; female flowers with tufts of long bristles

SHRUBS

Asteraceae - Aster Family

Baccharis halimifolia
Eastern Baccharis

National Wetland Plant List
Mtn/Pdmt: **FACW**   CP: **FAC**

Coefficient of Conservatism
Mtn: **0**   Pdmt: **1**   CP: **3**

Not native in Piedmont and Mountains; toxic leaves and seeds

Drawing: Karen Kendig

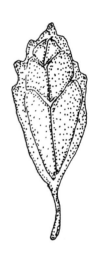

Habit
Broad-leaved deciduous to semi-evergreen woody shrub or small tree with ascending branches 1-4 m tall.

Leaves
Alternate and toothed or entire with serrations mostly towards leaf apex. Leaves elliptic to obovate, 3-7 cm long and 1-4 cm wide, conspicuously pale blue-green. Leaves just below flowers not serrated.

Flowers/Fruit
White feathery flowers in small heads arranged in stalked clusters at branch tips. Separate male and female plants. Blooms and fruits September to November.

Field Characteristics
Baccharis is the only member of the aster family in eastern North America to reach "tree" stature. When in flower and in fruit, the shrub appears white due to the cotton-like flowers.

Habitat/Range
Brackish marsh edges, ditch banks, old fields, damp thickets. Originally found just in the Coastal Plain, *Baccharis halimifolia* (eastern baccharis) has spread widely beyond its original range into the Piedmont, rare in the Mountains. It is considered non-native in the Piedmont and Mountains.

Similar Species
Can be confused with *Iva frutescens* (marsh elder) in coastal areas; see Common Confusions section, p. 403. Also called groundsel tree.

Borrichia frutescens — Sea Ox-Eye Daisy

Mature plant has woody stem, branching pattern

Succulent, light green leaves

Yellow, daisy-like flowers

Spiky, ball-shaped seedheads, brown at maturity

Can grow in dense colonies

SHRUBS

Asteraceae - Aster Family

Borrichia frutescens
Sea Ox-Eye Daisy

National Wetland Plant List
Mtn/Pdmt: **n/a**   CP: **OBL**

Coefficient of Conservatism
Mtn: **n/a**   Pdmt: **n/a**   CP: **8**

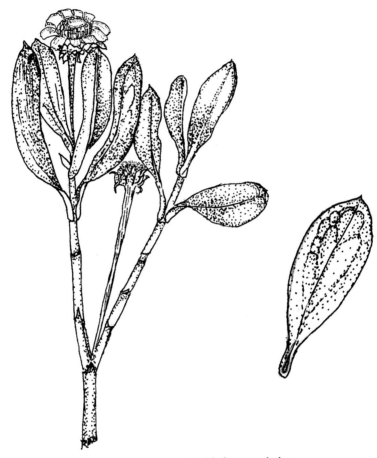

Drawing: Karen Kendig

Habit
Low-growing succulent, rhizome-forming shrub, often growing in extensive colonies almost to 1 m tall.

Leaves
Opposite, thick, narrow to broadly oblanceolate, 2-8 cm long and 1-3 cm wide. Leaves mostly entire, although sometimes slightly dentate. Petiole bases attached in a "U" shape against stem.

Flowers/Fruit
Attractive yellow flowers are typical of the Aster family. Spiky ball-shaped seedheads brown when mature. Blooms and fruits May to September.

Field Characteristics
Three prominent veins in leaves are distinctive as well as the "daisy" flowers, which persist from summer through fall.

Habitat/Range
Common at the edges of salt marshes and mud flats in the outer Coastal Plain.

Cephalanthus occidentalis — Common Buttonbush

Glossy, opposite leaves with brown twigs

Unique spherical pin-cushion flowers

Usually found in standing water

Spherical, "button" fruit

Brown twigs with raised elongated lenticels

SHRUBS

Rubiaceae - Madder Family

Cephalanthus occidentalis
Common Buttonbush

National Wetland Plant List
Mtn/Pdmt: **OBL**   CP: **OBL**

Coefficient of Conservatism
Mtn: **6**   Pdmt: **5**   CP: **5**

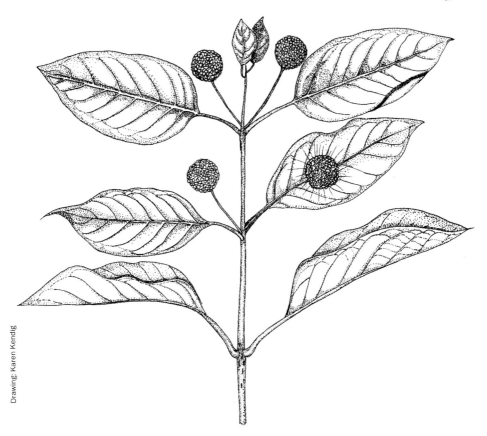

Drawing: Karen Kendig

Habit
Deciduous shrub, 1-3 m tall, usually in standing water.

Leaves
Broad, shiny leaves ovoid to elliptic, with pointed ends. Shiny, opposite or whorled and entire. Leaves 7-15 cm long, 3-10 cm broad.

Flowers/Fruit
Flowers are small white tubes, formed on spheres about 3 cm in diameter. Clusters appear as balls or "buttons" following summer flowering period. Flowers June to August; fruits August/September.

Field Characteristics
Brown twigs have raised elongated lenticels. A brown triangular membrane is present between petioles. Spherical "buttons" persist through winter. Early settlers were said to have used the buttons in clothing.

Habitat/Range
Usually in standing water at perimeters of lakes, ponds, freshwater marshes, forested wetlands, along streams throughout NC, but less frequent in the mountains.

Similar Species
Cornus amomum (silky dogwood) also has opposite, entire leaves, but with brown hairs beneath and thin fibers evident when leaves are broken.

Clethra alnifolia — Coastal Sweet-pepperbush

Flowers from all sides of flowering stem

Leaf wider past middle; lower part untoothed

Immature rounded fruits

Often found in colonies, especially in wet sands

Seed capsules persist through winter and spring

SHRUBS

Clethraceae - Pepperbush Family

Clethra alnifolia
Coastal Sweet-pepperbush

National Wetland Plant List
Mtn/Pdmt: **FAC** CP: **FACW**

Coefficient of Conservatism
Mtn: **n/a** Pdmt: **7** CP: **6**

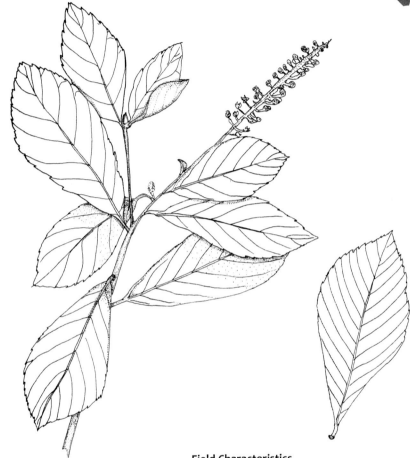

Drawing: Karen Kendig

Habit
Broad-leaved deciduous shrub to 3 m tall.

Leaves
Alternate, elliptic, obovate or oblanceolate leaves. Leaves toothed and approximately 8 cm long by 4 cm wide. Typically widest toward the tip.

Flowers/Fruit
Fragrant, small, white, 5-petaled flowers produced in racemes. Blooms June/July; fruits September/October.

Field Characteristics
Racemes with empty seed capsules persist through winter. Thick, light brown scales cover the current year's stems. Distinctive leaf shape.

Habitat/Range
Wet pine savannas, flatwoods, bays, and pocosins in the Coastal Plain, and damp floodplains of the lower Piedmont.

Similar Species
Similar to *Itea virginica* (Virginia sweetspire), but note leaf shape of *Clethra alnifolia* (coastal sweet-pepperbush) is widest past the middle, toward leaf tip. Teeth on *C. alnifolia* are absent from the base of the leaf, whereas in *Itea virginica*, teeth exist along the whole margin. See Common Confusions section, p. 404.

Cornus amomum — Silky Dogwood

Curved raised veins on leaf undersides

White "threads" visible in broken leaf (all dogwoods)

Fruits turn blue when mature

Brownish, curly, appressed hairs on leaf underside

Leafy shrub, to 5 m

SHRUBS

Cornaceae - Dogwood Family

Cornus amomum
Silky Dogwood

National Wetland Plant List
Mtn/Pdmt: **FACW** CP: **FACW**

Coefficient of Conservatism
Mtn: **5** Pdmt: **5** CP: **5**

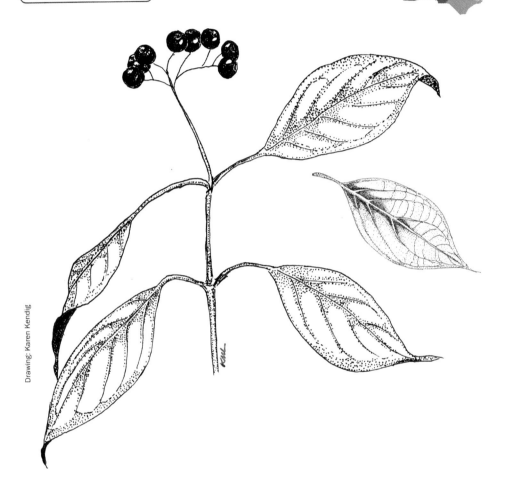

Drawing: Karen Kendig

Habit
Deciduous shrub reaching up to 5 m.

Leaves
Opposite, entire, ovate or elliptic-shaped leaves with typical dogwood venation and smooth margins. Leaves up to 10 cm long and 7 cm wide. Brownish appressed hairs on undersurface of leaves, especially on veins.

Flowers/Fruit
Flat-topped cyme, which produce blue drupes (berry-like fruits). Blooms May/June; fruits August/September.

Field Characteristics
In all dogwoods, white "threads" visible when leaf is broken and pulled apart.

Habitat/Range
Marshes, swamp forests, along rivers and streams mainly in the Piedmont and Mountains. Occasionally in the Coastal Plain.

Similar Species
Pith of second year growth is brown, whereas the pith of *Cornus foemina* (swamp dogwood - more coastal) is white. *Cephalanthus occidentalis* (common buttonbush) also has opposite entire leaves, but they lacks hairs beneath.

Cyrilla racemiflora — Swamp Titi

Flowers in long hanging racemes

Bark of older shrubs reddish

Raised midrib on leaf undersides

Leaves generally much wider past the middle

Glossy leaves somewhat variable in shape

SHRUBS

Cyrillaceae - Cyrilla Family

Cyrilla racemiflora
Swamp Titi

National Wetland Plant List
Mtn/Pdmt: **FACW**   CP: **FACW**

Coefficient of Conservatism
Mtn: **n/a**   Pdmt: **7**   CP: **6**

Drawing: Karen Kendig

Habit
Deciduous to semi-evergreen shrub or small tree to 8 m tall, often forming dense thickets.

Leaves
Alternate, usually narrow and obovate or elliptic, clustered near branch tips. Variable in size, up to 10 cm long and 2-3 cm wide and somewhat variable in shape. Glossy leaves have raised midribs beneath.

Flowers/Fruit
Racemes of white flowers from 5 to 15 cm long originate from the previous season's growth. Blooms May to July; fruits September/October.

Field Characteristics
Look for glossy leaves with pronounced midribs beneath. Distinctive raised ridge under leaf scar. Racemes of dry fruits persist through winter. Bark of older shrubs reddish.

Habitat/Range
Pocosins, swamps, pine flatwoods and streambanks in acidic, sandy or peaty soils, chiefly in the Coastal Plain and sometimes in the Piedmont.

Decodon verticillatus — Swamp Loosestrife

Clusters of purple flowers along stems

Shrub habit; reddish brown bark sloughs off

Long arching stems will root at tips

Stems become spongy in water

Fruit is dark brown capsules

SHRUBS

Lythraceae - Loosestrife Family

Decodon verticillatus
Swamp Loosestrife

National Wetland Plant List
Mtn/Pdmt: **OBL**   CP: **OBL**

Coefficient of Conservatism
Mtn: **8**   Pdmt: **7**   CP: **7**

Drawing: Sara Fish Brown/E.O. Beal

Habit
Shrubby perennial with long arching leafy stems, some rooting at tips. Can also be considered herbaceous.

Leaves
Opposite or whorled, lance-shaped, large, to 20 cm long and 5 cm wide, with a prominent white or pink midvein. Leaves smooth above, pubescent with short hairs below.

Flowers/Fruit
Purple flowers about 3 cm wide, in clusters at leaf bases, with long, extending stamens. Blooms and fruits July to September.

Field Characteristics
Stems spongy below water; bark of stems above water sloughs off in long cinnamon-colored strips.

Habitat/Range
Shallow water around lakes and ponds, in marshes, swamps, and wet shrubby thickets in the Coastal Plain.

Taxonomic Note
This species is also called water-willow, but is more closely related to loosestrifes than willows.

Eubotrys racemosus — Swamp Fetterbush

Major veins loop back inward before outer edge; margins unevenly serrated

Dry, capsule fruit splits in 5 sections

Immature capsule fruit

Pointed flower buds

Flowers late March to early June

SHRUBS

Ericaceae - Heath Family

National Wetland Plant List
Mtn/Pdmt: **FACW**   CP: **FACW**

Coefficient of Conservatism
Mtn: **9**   Pdmt: **8**   CP: **7**

Eubotrys racemosus
Swamp Fetterbush

Drawing: Karen Kendig

Habit
Deciduous shrub reaching to 4 m, but typically smaller.

Leaves
Alternate, elliptic, finely (but unevenly) serrated leaves. Size varies greatly on branches. Leaves 3-9 cm long, 1-4 cm wide.

Flowers/Fruit
Sharply pointed flower buds develop in summer, opening the following spring. White "urn" shaped flowers on straight or slightly arching racemes, as long as 9 cm. Fruit are dry, brown capsules with 5 sutures and prominent styles remaining from flower. Blooms late March to early June; fruits September/October.

Field Characteristics
Distinctive fruit capsules in racemes. Difficult to identify without flowers/fruits; however, in summer, this species develops flower buds that remain on twigs all winter, until opening in spring.

Habitat/Range
Swamps, cypress-gum depressions, along shorelines, mainly in Piedmont and Coastal Plain in forested wetlands with standing water.

Similar Species
Itea virginica (Virginia sweetspire) has similar leaves, but its flower clusters/capsules are on all sides of the stem. Main leaf veins of *Eubotrys racemosus* (swamp fetterbush) all curve back toward mid-vein, whereas in *Itea virginica*, lower veins extend to leaf edge. See Common Confusions section, p. 404. Also similar to *Leucothoe axillaris* (coastal doghobble), which has branched racemes and is evergreen.

Taxonomic Note
Synonym: *Leucothoe racemosa*

Euonymus americanus — American Strawberry-bush

Leaves opposite along green twigs

Flat, yellowish, 5-petaled flowers at leaf axils

Immature fruits on long stems

Mature pink fruits split to reveal orange seeds

Mature bright pink-red fruits are bumpy, lobed capsules

SHRUBS

Celastraceae - Bittersweet Family

Euonymus americanus

American Strawberry-bush

National Wetland Plant List
Mtn/Pdmt: **FAC**   CP: **FAC**

Coefficient of Conservatism
Mtn: **5**   Pdmt: **5**   CP: **5**

Toxic Plant

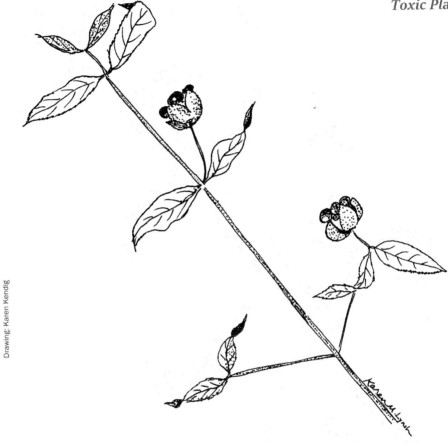

Drawing: Karen Kendig

Habit
Small deciduous shrub to 2 or 3 m, but most commonly 1 m.

Leaves
Opposite, serrated, lance-shaped leaves., in sets of two along twigs. Deciduous leaves slow to drop, sometimes lasting until early winter.

Flowers/Fruit
Small, light green, 5-petaled flowers. Fruit is bumpy, lobed, reddish capsule which splits into 3-5 sections, exposing orange or red seeds. Blooms May/June; fruits September/October.

Field Characteristics
Green 4-sided branches and attractive fruits distinctive. Fruits appear like strawberries or bursting hearts; toxic if eaten in large quantities.

Habitat/Range
Stream banks, slopes, rich woodlands throughout NC.

Taxonomic Note
Also known as Bursting-heart and Hearts-a-busting.

Gaylussacia frondosa — Blue Huckleberry

Generally low growing shrub with light green leaves

Green berries mature to dark blue

Blueberry-like fruits are edible

Yellow glands on leaf undersides

Greenish-white to pinkish urn-shaped flowers

SHRUBS

Ericaceae - Heath Family

National Wetland Plant List
Mtn/Pdmt: **FAC**   CP: **FAC**

Coefficient of Conservatism
Mtn: **n/a**   Pdmt: **7**   CP: **6**

Gaylussacia frondosa
Blue Huckleberry

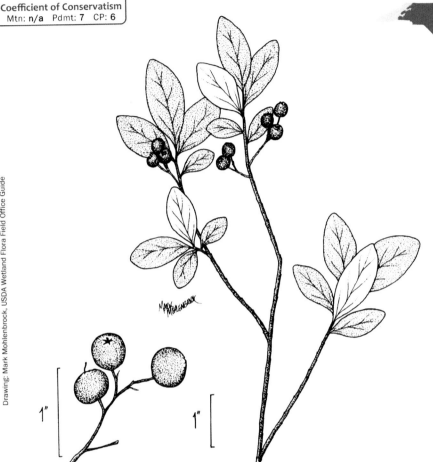

Habit
Deciduous, low growing shrub, usually 1 m tall or less.

Leaves
Alternate, simple, oval shaped, entire leaves with short petioles. Leaves pale grayish-green beneath, can be smooth or pubescent. Tiny yellow resinous dots only on underside.

Flowers/Fruit
Racemes with greenish-white to pinkish urn-shaped flowers and usually 2 branchlets. Edible berries are green, ripening to dark blue or black. Blooms late March to May; fruits June to August.

Field Characteristics
Rubbing the leaf undersides on paper will turn it yellow, from the yellow resinous glands. Berries are edible.

Habitat/Range
Moist acidic woodlands, especially sandhill pocosins and pine savanna-pocosin edges. Uncommon in the Piedmont, but common in the Coastal Plain, especially southeast.

Similar Species
Vaccinium species (blueberries) lack the yellow resinous dots on leaves.

Hibiscus moscheutos — Swamp Rose Mallow

Large, showy, white flowers

Ovate leaves with pointed tips; some with side lobes

Robust shrub, stems rising from base

Clusters of buds at branch tips

Dry fruit capsules split at maturity

SHRUBS

Malvaceae - Mallow Family

National Wetland Plant List
Mtn/Pdmt: **OBL** CP: **OBL**

Coefficient of Conservatism
Mtn: **5** Pdmt: **5** CP: **5**

Hibiscus moscheutos
Swamp Rose Mallow

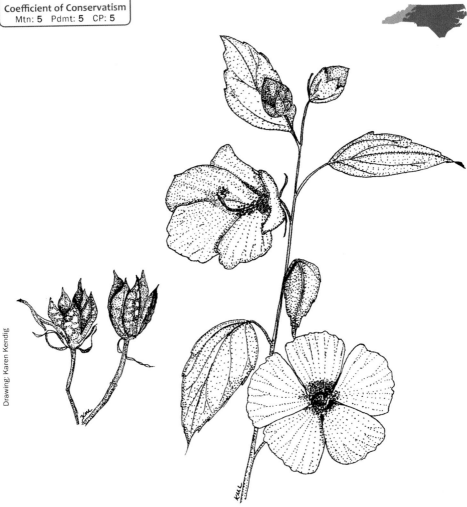

Habit
Tall perennial woody shrub to 2 m, with stems rising from base. Upper stems hairy. Deciduous.

Leaves
Alternate, pubescent, leaves are oval or 3-lobed and toothed.

Flowers/Fruit
Large, showy, creamy-white, 5-petaled flowers with a crimson center, occasionally pinkish. Flowers about 20 cm wide. Fruit capsule splits into 5 parts and persists through winter. Blooms June to September; fruits July to October.

Field Characteristics
Large, showy, white flowers; fruit capsules persisting through winter.

Habitat/Range
Fresh to slightly brackish marshes throughout the state; most abundant in the Piedmont and Coastal Plain.

Similar Species
Similar to *Kosteletzkya virginica* (saltmarsh mallow), but the *K. virginica* flower is smaller, pink, with a much longer pistil. Leaves of *K. virginica* are all lobed.

Hypericum hypericoides — St. Andrew's-Cross

A small, leafy understory shrub

Small linear leaves on dark brown stems

Four yellow petals in X shape, with wide green sepals

Flat, leaf-like fruits contain seeds

Thin, dark brown stems; oval, brown seed capsules

SHRUBS

Hypericaceae - St. John's Wort Family

Hypericum hypericoides
St. Andrew's-Cross

National Wetland Plant List
Mtn/Pdmt: **FACU**   CP: **FAC**

Coefficient of Conservatism
Mtn: **5**   Pdmt: **5**   CP: **5**

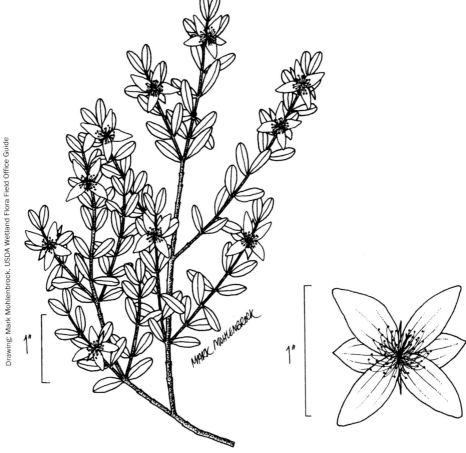

Drawing: Mark Mohlenbrock, USDA Wetland Flora Field Office Guide

Habit
Upright, semi-evergreen to deciduous small woody shrub, to 1 m tall.

Leaves
Leaves are opposite, narrow, less than 2.5 cm long, tapered at the base.

Flowers/Fruit
Yellow flowers with four narrow petals, arranged like a capital X, born singly or in small groups on twig ends. Two prominent sepals below. Flat, oval, brown seed capsules. Flowering and fruiting May to August.

Field Characteristics
Usually grows as a very small shrub, with fine, dark twigs. Small, narrow leaves fairly distinctive.

Habitat/Range
Common in forests throughout the eastern half of the Piedmont and all of the Coastal Plain; uncommon in the Mountains.

Ilex coriacea — Large Gallberry

Upright shrub, to 5m tall

Small white flowers April/May

Somewhat flattened, globular drupe fruits

Fruits ripen September/October

Rounded, thick leaf with scattered short spines on upper half; small black stipule at leaf base

SHRUBS

Aquifoliaceae - Holly Family

Ilex coriacea
Large Gallberry

National Wetland Plant List
Mtn/Pdmt: **FACW**   CP: **FACW**

Coefficient of Conservatism
Mtn: **n/a**   Pdmt: **8**   CP: **7**

Drawing: Karen Kendig

Habit
Upright, evergreen shrub, to 5 m tall.

Leaves
Alternate, leathery, elliptic or obovate leaves, with scattered short spines along the upper half of the leaf margin. Leaves 4 to 9 cm long and 1.5 to 4 cm wide.

Flowers/Fruit
Fruit maturing from red to shiny black, somewhat flattened, globular drupe which drops off when mature. Blooms April/May; fruits ripen September/October.

Field Characteristics
Thick rounded leaves with scattered short spines easily felt. All hollies have small black stipules at leaf bases.

Habitat/Range
Acidic Coastal Plain wetlands such as pocosins, bay forests, pine wetlands, and stream banks. Often forms large colonies.

Similar Species
Similar to *Ilex glabra* (inkberry), but generally a taller plant with wider leaves. Short spines more easily felt along edges than those of *Ilex glabra*. Fruits larger than those of *Ilex glabra* and somewhat flattened, rather than globular.

Ilex decidua — Possumhaw

A large possumhaw

Gray twigs and limbs have small corky bumps

Small leaves, alternately arranged

Leaves soft and thin, without spines

Flowers and fruit on short stalks at bases of leaves; corky leaf scars on twigs distinctive

SHRUBS

Aquifoliaceae - Holly Family

Ilex decidua
Possumhaw

National Wetland Plant List
Mtn/Pdmt: **FACW**   CP: **FACW**

Coefficient of Conservatism
Mtn: **n/a**   Pdmt: **6**   CP: **6**

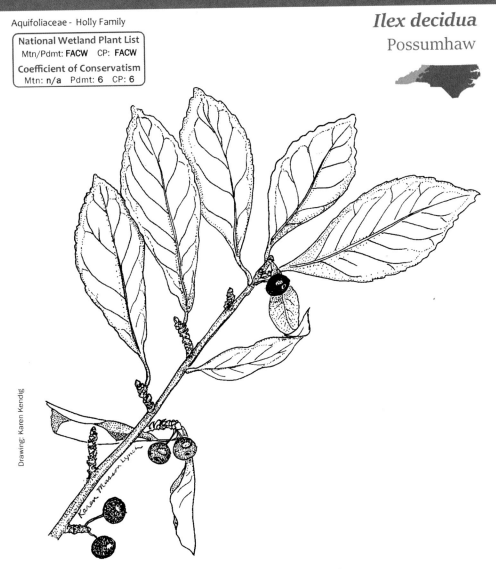

Drawing: Karen Kendig

Habit
Large deciduous shrub or small understory tree, reaching 10 m.

Leaves
Elliptic to obovate, with crenate margins, 3-5 cm long and 1-3 cm wide, with the widest part past the middle. Leaves soft, not leathery, and without spines.

Flowers/Fruit
Small white flowers on short stalks at bases of leaves. Fruit is red spherical drupe that persists after leaves fall. Blooms March to May; fruits September/October.

Field Characteristics
Deciduous leaves on very short shoots with many leaf scars. Corky leaf scars on gray twigs distinctive. All hollies have small black stipules at leaf bases.

Habitat/Range
Floodplain forests and along creeks and uplands. Chiefly in the Piedmont and northwestern third of the Coastal Plain.

Similar Species
Ilex verticillata (common winterberry) has similar, but larger and wider, leaves.

Ilex glabra — Inkberry

Evergreen shrub, to 2 or 3 m tall

Fruits mature green to black; September to November

Spines point inward toward leaf

Often in colonies; mainly in the Coastal Plain

Small, white flowers May/June

SHRUBS

Aquifoliaceae - Holly Family

Ilex glabra
Inkberry

National Wetland Plant List
Mtn/Pdmt: **FAC**   CP: **FACW**

Coefficient of Conservatism
Mtn: **n/a**   Pdmt: **6**   CP: **5**

Habit
Evergreen, colonial shrub to 2 or 3 m tall.

Leaves
Elliptic to obovate or oblanceolate leathery green leaves. Leaves 2-5 cm long and 1-2 cm wide. Leaves pointed at tips and have a pair or two of teeth pointing toward the apex. Undersides of leaves contain scattered, punctate, dark or reddish glands.

Flowers/Fruit
Fruit is a spherical or slightly flattened berry-like drupe, green maturing to black. Blooms May/June; fruits September to November.

Field Characteristics
Short spines, if present, are not prominent and pointed inward toward leaf apex. Fruits (drupes) persistent throughout winter. Stipule at leaf base dark brown and obvious.

Habitat/Range
Abundant in acidic pine wetlands, pocosins, bay forests. Mainly a Coastal Plain species.

Similar Species
Leathery, evergreen leaves narrower than those of *Ilex coriacea* (large gallberry), with less obvious spines.

Taxonomic Note
Also known as gallberry.

Ilex verticillata — Common Winterberry

Flowers on short stalks at bases of leaves

Green berries mature to bright red

Branching veins deeply impressed into serrated leaves

Smooth, gray bark with white lenticels

Flower/fruit clusters emerge from leaf bases

SHRUBS

Aquifoliaceae - Holly Family

National Wetland Plant List
Mtn/Pdmt: **FACW**   CP: **FACW**

Coefficient of Conservatism
Mtn: **7**   Pdmt: **7**   CP: **7**

Ilex verticillata
Common Winterberry

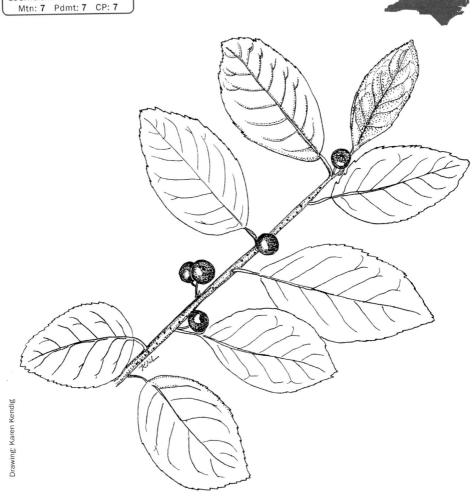

Drawing: Karen Kendig

Habit
Large deciduous shrub ranging from 1-4 m.

Leaves
Elliptic to obovate 4-10 cm long and 2-5 cm wide. Margins serrated and leaf surfaces rough (pubescent) with depressed veins.

Flowers/Fruit
The fruit is a drupe which appears as a bright red, spherical berry. Blooms April/May; fruits September to November. Fruit may persist into winter.

Field Characteristics
Bumpy lenticels on branches and striking, red, berry-like drupes distinctive. All hollies have small black stipules at leaf bases.

Habitat/Range
Swamps, alongside streams and wet woodlands, scattered across the state, but less common in the eastern Coastal Plain.

Similar Species
Ilex decidua (possumhaw) has similar, but smaller and narrower, leaves, as well as dense leaf scars on twigs.

Itea virginica — Virginia Sweetspire

Small white flowers in long racemes

Prominent major veins, minor veins connecting them

Fruits in fuzzy capsules

Teeth with tiny spines around whole leaf margin

Deciduous, sprawling shrub, to 2 m

SHRUBS

Grossulariaceae - Currant Family

National Wetland Plant List
Mtn/Pdmt: **OBL**   CP: **FACW**

Coefficient of Conservatism
Mtn: **7**   Pdmt: **7**   CP: **7**

Itea virginica
Virginia Sweetspire

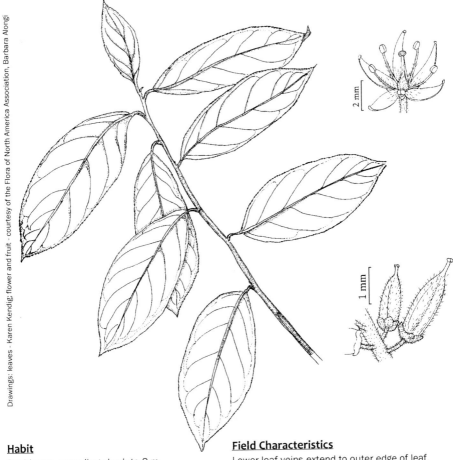

Drawings: leaves - Karen Kendig; flower and fruit - courtesy of the Flora of North America Association, Barbara Alongi

Habit
Deciduous, sprawling shrub to 2 m.

Leaves
Alternate, finely toothed, glabrous, elliptic leaves. Leaves oblong with parallel sides, 2-9 cm long and 1-4 cm wide.

Flowers/Fruit
Clusters of white, 5-petaled flowers form a narrow raceme at branch ends. Two-lobed beaked capsules persist. Flowers May/June; fruits soon after.

Field Characteristics
Lower leaf veins extend to outer edge of leaf. Branches often green above and burgundy on undersides. Stems have a white chambered pith.

Habitat/Range
Low woods, swamps, alongside streams, throughout NC.

Similar Species
Similar to *Clethra alnifolia* (coastal sweet-pepperbush), but leaves thinner and teeth extend all the way around the leaf margin, whereas in *C. alnifolia*, they are absent from leaf bases. *Eubotrys racemosus* (swamp fetterbush) leaves have all major veins curving inward before reaching leaf margins. See Common Confusions section, p. 404.

Iva frutescens — Marsh Elder

Dense woody shrub with many branches

Upper leaves untoothed

Leaves somewhat succulent

Iva frutescens (left) vs *Baccharis halimifolia* (right)

Often grows in colonies

SHRUBS

Asteraceae - Aster Family

Iva frutescens
Marsh Elder

National Wetland Plant List
Mtn/Pdmt: **n/a**   CP: **FACW**

Coefficient of Conservatism
Mtn: **n/a**   Pdmt: **n/a**   CP: **6**

Habit
Dense shrub with many branches, 1-2 m tall. Tardily deciduous to nearly evergreen.

Leaves
Opposite (except in flowering branchlets), elliptic to lance-shaped, fleshy leaves, 3 to 8 cm long and 0.5 to 2 cm wide.

Flowers/Fruit
Small greenish flowers occur in terminal spikes 3-10 cm long. Flowers and fruits late August to November.

Field Characteristics
Terminal spikes of fruits with dark brown nutlets persist through winter.

Habitat/Range
Brackish marshes, estuarine shores, mud flats and vacant lots in the outer Coastal Plain.

Similar Species
Compare to *Baccharis halimifolia* (Eastern baccharis), which grows in similar habitats, but has wider, shorter, toothed leaves and bristly flowers only in fall (see leaf comparison photo, middle opposite). See Common Confusions section, p. 403.

Kosteletzkya virginica — Saltmarsh Mallow

Large, pink, showy flowers with long stigma stalk

Coarsely toothed and roughly pubescent leaves

Dry, splitting capsules

Tall plant with pink flowers, in tidal wetlands

Fruits soon after blooming July to October

SHRUBS

Malvaceae - Mallow Family

Kosteletzkya virginica
Saltmarsh Mallow

National Wetland Plant List
Mtn/Pdmt: **n/a**   CP: **OBL**

Coefficient of Conservatism
Mtn: **n/a**   Pdmt: **n/a**   CP: **6**

Drawing: Karen Kendig

Habit
Medium height deciduous shrub, 1-2 m with few to numerous branches.

Leaves
Alternate, coarsely toothed and roughly pubescent. Leaves sagittate or triangular in shape, with bottom leaves the largest and upper leaves reduced.

Flowers/Fruit
Pink, 5-petaled flower stemming from leaf bases. Flower 5-8 cm wide. Blooms July to October, fruiting soon after.

Field Characteristics
In tidally influenced wetlands, look for a tall plant with pink flowers and lobed leaves.

Habitat/Range
Brackish to rarely fresh tidal marshes, shores and ditches, swamps, wet woodlands in the outer Coastal Plain.

Similar Species
Similar to *Hibiscus moscheutos* (swamp rose mallow), although the flower of *H. moscheutos* is larger, white, and with a much shorter stigma stalk. *H. moscheutos* grows strictly in freshwater wetlands. Leaves of *Kosteletzkya virginica* (saltmarsh mallow) are always lobed.

Taxonomic Note
Synonym: *Kosteletzkya pentacarpos*

Leucothoe axillaris — Coastal Doghobble

Deeply veined leaves with pink petioles

White, urn-shaped flowers in branching racemes

Leafy, evergreen shrub with loose arching branches

Dry capsules split along 5 lines

Can grow in low colonies

SHRUBS

Ericaceae - Heath Family

National Wetland Plant List
Mtn/Pdmt: **FACW**   CP: **FACW**

Coefficient of Conservatism
Mtn: **n/a**   Pdmt: **7**   CP: **7**

Leucothoe axillaris
Coastal Doghobble

Highly Toxic Plant

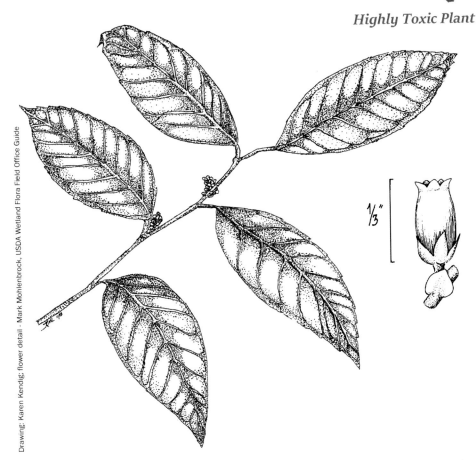

Habit
Low, evergreen shrub with loose, arching branches, up to 1.5 m tall.

Leaves
Alternate, lance-shaped to elliptic-shaped leaves with toothed margins. Leaves 5-13 cm long, 2-5 cm broad.

Flowers/Fruit
Whitish, urn-shaped flowers in branching racemes, originating from the axis and typically consist of more than 15 flowers. Fruits dry splitting, 5-lobed capsules. Flowers late March to May; fruits August to October.

Field Characteristics
Trailing branches with alternate evergreen leaves, usually in knee-high colonies in floodplain forests. Thick evergreen stands of this plant were said to make hunting dogs hobble, hence the common name. Flowers and leaves are highly toxic.

Habitat/Range
Wet, acidic swamps and depressions, bay forests, pocosins, seepages, mainly in the Coastal Plain.

Similar Species
Similar to *Eubotrys racemosus* (swamp fetterbush), which has flowers/fruits on unbranched racemes and is deciduous.

Ligustrum sinense — Chinese Privet

Evergreen shrub or small tree, to 10 m

White flowers in clusters at branch tips, May to June

Often grows in colonies, in the understory of damp forests

Opposite leaves and branches

Fruits mature to bluish-black drupes

SHRUBS

Oleaceae - Olive Family

National Wetland Plant List
Mtn/Pdmt: **FACU**   CP: **FAC**

Coefficient of Conservatism
Mtn: **0**   Pdmt: **0**   CP: **0**

Ligustrum sinense
Chinese Privet

Non-native; toxic plant

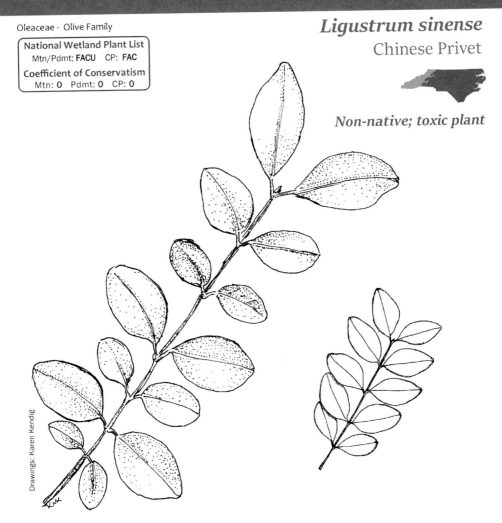

Drawings: Karen Kendig

Habit
Evergreen shrub or small tree, often forming dense colonies, to 10 m in height. Twigs and branchlets densely pubescent.

Leaves
Opposite, entire, elliptic or ovate leaves, 4 cm long and 1-2 cm wide. Leaf surfaces somewhat dull.

Flowers/Fruit
Small, white, (unpleasantly) fragrant flowers forming panicles. Flowers May through June, fruiting soon after. Fruits bluish-black drupes at maturity.

Field Characteristics
Opposite branches. Once established, this non-native rapidly colonizes waste areas and wetlands, to the exclusion of many other species. Toxic plant.

Habitat/Range
Low woods, disturbed wetlands, moist roadsides, floodplain forests, and waste areas throughout the state. One of the worst invasives in the state, along with *Lonicera japonica* (Japanese honeysuckle) and *Microstegium vimineum* (Japanese stilt grass), shading out and eliminating native flora.

Similar Species
Ligustrum sinense (Chinese privet) leaves are much smaller and less leathery than the taller *Ligustrum lucidum* (glossy privet), a nonnative which is also highly invasive.

Lindera benzoin — Northern Spicebush

Distinctive raised lenticels

Leaves feel, and sometimes appear, waxy/sticky coated

Thin, obovate leaves

Small, yellow flowers before leaves emerge

Suzanne Cadwell, inaturalist

Small, elliptical fruits mature to bright red

SHRUBS

Lauraceae - Laurel Family

National Wetland Plant List
Mtn/Pdmt: **FAC** CP: **FACW**

Coefficient of Conservatism
Mtn: **6** Pdmt: **6** CP: **6**

Lindera benzoin
Northern Spicebush

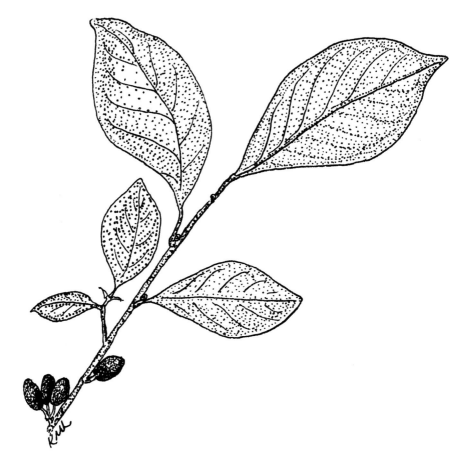

Drawing: Karen Kendig

Habit
Deciduous understory shrub to 3 m tall, often forming colonies alongside streams and in bottomlands.

Leaves
Alternate, obovate, thin, 6-14 cm long and 2-6 cm wide with acuminate tips and entire margins. Leaf undersides distinctly light green.

Flowers/Fruit
Small, yellow flowers which produce red, elliptical drupes up to 1 cm wide. Flowers March/April before leaves emerge; fruits August/September.

Field Characteristics
Leaves and twigs lemon-fragrant when crushed and the thin leaves have a slightly waxy or sticky coating. Stems have raised lenticels.

Habitat/Range
Stream margins and rich moist woods, floodplain forests, mainly in the Mountains, Piedmont, and brownwater river floodplains in the Coastal Plain.

Similar Species
Nyssa sylvatica (blackgum) is a small to medium tree that has similar oval, alternate leaves, but they lack the lemon scent and waxy or sticky coating. *N. sylvatica* fruits are larger and blue-black.

Lyonia ligustrina — Maleberry

Medium sized shrub, to 4 m

Leafy raceme with white, globular flowers

Fuzzy, round capsules split at maturity

White flowers with fused petals

Finely serrated leaves with short hairs on both surfaces; leaf shape somewhat variable

SHRUBS

Ericaceae - Heath Family

Lyonia ligustrina
Maleberry

National Wetland Plant List
Mtn/Pdmt: **FACW**   CP: **FACW**

Coefficient of Conservatism
Mtn: **7**   Pdmt: **7**   CP: **7**

Habit
Semi-evergreen or deciduous, medium sized shrub, to 4 m.

Leaves
Leaves alternate, ovate, finely serrated with short hairs on both surfaces; smaller leaves in wetter habitats.

Flowers/Fruit
Leafy raceme with white, globular flowers with fused petals; flowers produce fuzzy, round capsules that split when ripe. Flowers late April to June; fruits September/October.

Field Characteristics
Young twigs have red bark. Inflorescence is leafy.

Habitat/Range
Pocosins, seepage wetlands, mountain bogs, bottomlands, savannas, and pine flatwoods. Found throughout the state, but most abundant in the Coastal Plain.

Similar Species
Called maleberry because it is similar to *Vaccinium* spp. (blueberry) or *Gaylussacia* spp. (huckleberry), but produces dry capsules and not fleshy edible fruit.

Taxonomic Note
Mountains and Piedmont variety is var. *ligustrina*. Leaves appear gray-green in the Coastal Plain variety (var. *foliosiflora*).

Lyonia lucida — Fetterbush Lyonia

Evergreen shrub, usually 1 m but can reach 3 m

Immature fruits; prominent midribs on leaves

Fruits mature to dry brown capsules that split

Pinkish urn-shaped flowers

Distinctive thick, rounded leaves with ribbed edges

SHRUBS

Ericaceae - Heath Family

National Wetland Plant List
Mtn/Pdmt: **FACW**   CP: **FACW**

Coefficient of Conservatism
Mtn: **0**   Pdmt: **7**   CP: **7**

Lyonia lucida
Fetterbush Lyonia

Highly Toxic Plant

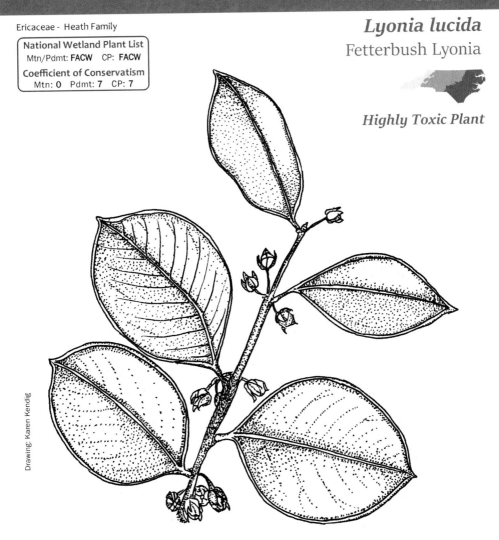

Drawing: Karen Kendig

Habit
Small evergreen shrub, although may reach as high as 3 m. Usually colonial in nature.

Leaves
Leathery, dark green leaves with very smooth, entire margins with marginal veins. Leaves generally elliptic, 3-9 cm long by 1-4 cm wide on flattened branches.

Flowers/Fruit
Beautiful when flowering, with clusters of pinkish (sometimes white), urn-shaped flowers arising from axis. Fruit an oval capsule, about 5 mm long. Blooms April to early June; fruits September/October.

Field Characteristics
Note thick, shiny, evergreen leaves with prominent mid-vein. Veins also encircling leaf margins very characteristic. Branches distinctly flattened below nodes.

Habitat/Range
Usually found where surface water is present most of the year in pocosins, blackwater swamp forests, pine wetlands, and bay forests, mainly in the Coastal Plain.

Morella caroliniensis — Southern Bayberry

Medium sized evergreen shrub to 2 m high and wide

Flowers look like short catkins; black or brown twigs

Wider, larger leaves than wax myrtle (*Morella cerifera*)

Bluish-black or dark brown, globulose fruits

Yellow, resinous dots only on backside of leaves

SHRUBS

Myricaceae - Bayberry Family

Morella caroliniensis
Southern Bayberry

National Wetland Plant List
Mtn/Pdmt: **FAC**   CP: **FACW**

Coefficient of Conservatism
Mtn: **n/a**   Pdmt: **7**   CP: **7**

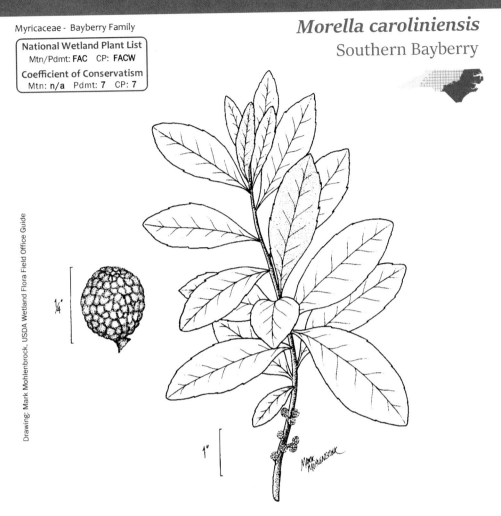

Habit
Medium sized evergreen shrub to 2 m.

Leaves
Alternate, elliptic, oblanceolate or obovate leathery leaves with a few shallow serrations toward tips of thick leaves. Yellow, resinous glands on the leaf underside only.

Flowers/Fruit
Male and female flowers borne on separate plants. Bluish or dark brown globose fruits may appear white from waxy covering. Blooms mainly in April; fruiting August to October.

Field Characteristics
Leaves only slightly aromatic when crushed; yellow, resinous dots only on undersides of leaves. Black or brown twigs.

Habitat/Range
Pocosins, bay forests, wet pine savannas and flatwoods in the Coastal Plain; rarely in the Piedmont.

Similar Species
Similar to the more abundant *Morella cerifera* (common wax myrtle), but *M. caroliniensis* has broader, thicker leaves with yellow waxy dots only on the leaf underside. Leaves are thicker and usually wider than *M. cerifera* (southern bayberry) and much less aromatic.

Taxonomic Note
Synonym: *Myrica heterophylla*

Morella cerifera — Common Wax Myrtle

Yellow, resinous dots on both sides of leaves

Female flowers look like fat catkins

The only *Morella* species reaching small tree heights, to 7 m

Smooth, gray bark

Small, waxy, whitish or gray fruits

SHRUBS

Myricaceae - Bayberry Family

Morella cerifera
Common Wax Myrtle

National Wetland Plant List
Mtn/Pdmt: **FAC**   CP: **FAC**

Coefficient of Conservatism
Mtn: **n/a**   Pdmt: **5**   CP: **4**

Drawing: Courtesy of the Flora of North America Association, John Myers

Habit
Evergreen large shrub or small tree to 7 m.

Leaves
Alternate, narrowly elliptic, oblanceolate leaves with a few serrations from the middle of leaves toward leaf tips. Leaves always less than 2.5 cm wide, and have yellow resinous glands on both sides of leaves.

Flowers/Fruit
Male and female flowers on separate shrubs. Male shrubs produce catkins in spring and female shrubs have oval, white or gray waxy fruits. Blooms mainly in April; fruiting August to October.

Field Characteristics
Leaves very fragrant when crushed; resinous dots on both leaf surfaces. Berries used for scents and candles.

Habitat/Range
Low ground, pond edges, pine wetlands, swamps, brackish marshes, and other moist habitats. Chiefly in the Coastal Plain but common in the Piedmont.

Similar Species
A similar species, *Morella caroliniensis* (southern bayberry) has larger, wider leaves and resinous dots on leaf underside only, not both surfaces. *M. cerifera* (common wax myrtle) is the only *Morella* species in the state reaching small tree height.

Taxonomic Note
Synonym: *Myrica cerifera*

Rosa palustris — Swamp Rose

Showy pink flowers bloom singly

Deciduous, thorny shrub, to 2m

Finely serrated, compound leaves

Rosehips covered with short bristles

Long stipules wrap around bases of petioles

SHRUBS

Rosaceae - Rose Family

Rosa palustris
Swamp Rose

National Wetland Plant List
Mtn/Pdmt: **OBL**   CP: **OBL**

Coefficient of Conservatism
Mtn: 6   Pdmt: 6   CP: 6

Drawing: Karen Kendig

Habit
Broad-leaved, deciduous shrub with decurved, thorn-like prickles. Grows to 2 m tall and reproduces by runners, sometimes forming thick stands.

Leaves
Alternate, pinnately compound. Leaflets elliptic and finely toothed. Number varies from 5-9 leaflets, usually 7. Leaflets 1-5 cm long and 0.5 to 2 cm wide with largest leaflets toward tip of leaf.

Flowers/Fruit
Large, pink, 5-petaled, single flowers at branch tips, later forming red rosehips. Blooms May to July; fruits September/October.

Field Characteristics
Recognized by its typical rose features (prickles, rosehips). Note long stipules wrapped at bases of petioles and bristles on rosehips.

Habitat/Range
Marshes or wet shores of streams, lakes, and swamps throughout NC, except rare in the Sandhills area.

Similar Species
A similar species, *Rosa multiflora* (multiflora rose) has clusters of several small, white fragrant flowers instead of single, large, pink flowers. It also has stipules that are flat or open with hairs on the edges. *R. multiflora* is non-native and usually grows in uplands.

Rubus pensilvanicus   Sawtooth Blackberry

Showy, white, 5-petaled flowers

Flowers at branch ends (racemes, solitary, or panicles)

Stalks last two years (branching and flowering second year)

Edible berries May to July

Compound leaves with three leaflets, serrated edges, deep veins, hairy beneath

SHRUBS

Rosaceae - Rose Family

Rubus pensilvanicus
Sawtooth Blackberry

National Wetland Plant List
Mtn/Pdmt: **FAC**   CP: **FAC**

Coefficient of Conservatism
Mtn: **3**   Pdmt: **2**   CP: **2**

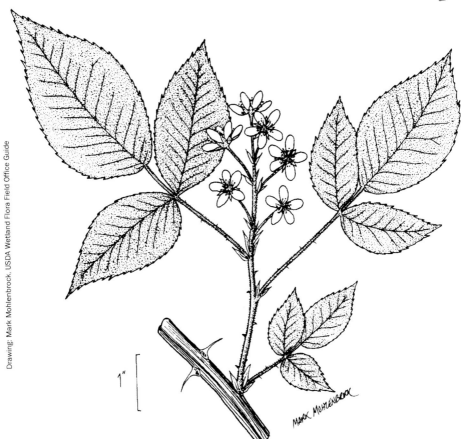

Habit
Thorny shrub with arching or erect branches, often forming extensive colonies.

Leaves
Leaves palmately compound, usually in 3 leaflets, with serrated edges and deep veins, hairy beneath.

Flowers/Fruit
White, usually 5-petaled, showy flowers at branch ends; may be solitary, in racemes, or in panicles. Flowers with numerous stamens and pistils. Fruit is a juicy, black, aggregated drupe, globular or elongated. Blooms April/May; fruits May to July.

Field Characteristics
Old, dead stalks brown and brittle but still thorny. Stalks last two years: first year - long stalks rise unbranched from base; second year - stalks branch and flower. Berries are edible.

Habitat/Range
Poorly drained fields, depressions, swales, pocosins, edges of marshes and swamps, swales, bogs, disturbed areas. Found at lower elevations in Mountains.

Taxonomic Note
The former *Rubus argutus* is now included within *R. pensilvanicus*.

Salix caroliniana — Coastal Plain Willow

Leaves whitish with prominent midvein beneath

Male catkins

Rounded shrub form

Usually grows as many-trunked shrub

Long, flowing leaves, finely toothed

SHRUBS

Salicaceae - Willow Family

National Wetland Plant List
Mtn/Pdmt: **OBL**   CP: **OBL**

Coefficient of Conservatism
Mtn: **n/a**   Pdmt: **4**   CP: **4**

Salix caroliniana
Coastal Plain Willow

Habit
Deciduous, large shrub more than small tree.

Leaves
Alternate, lance-shaped or sickle shaped leaves, finely toothed. Leaf blades whitish beneath. Stipules usually obvious at base of leaves.

Flowers/Fruit
Male and female catkins on separate trees. Fruits 1 cm, brown, flask-shaped, and crowded in long clusters. Flowers and fruits March/April.

Field Characteristics
Slender, reddish-brown twigs; leaves whitish or gray-green beneath.

Habitat/Range
Riverbanks, sandbars, and edges of ponds and lakes.

Similar Species
Salix nigra (black willow) has leaves greenish beneath, usually growing as a few-trunked small tree. *S. caroliniana* (Coastal Plain willow) has slightly smaller leaves, whitish beneath, usually growing as a multi-stemmed large shrub.

Sambucus nigra — Black Elderberry

Large, dense inflorescences of small, white flowers

Twigs have raised lenticels

Berries can be red to blue-black

Deciduous shrub to 4 m tall; overall rounded form

Opposite, compound leaves with 5 to 11 leaflets; leaflets finely toothed

SHRUBS

Adoxaceae - Moschatel Family

National Wetland Plant List
Mtn/Pdmt: **FAC** CP: **FACW**

Coefficient of Conservatism
Mtn: **4** Pdmt: **3** CP: **3**

Sambucus nigra
Black Elderberry

Toxic Plant

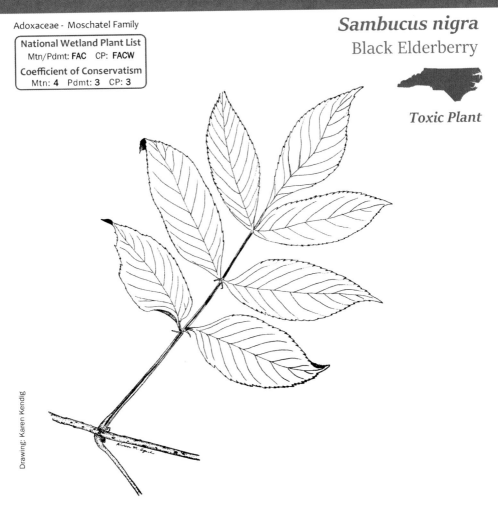

Drawing: Karen Kendig

Habit
Deciduous shrub to 4 m tall, stems with white, spongy or hollow pith.

Leaves
Opposite and pinnately compound with 5-11 leaflets (usually 7). Leaflets with toothed margins. Lower leaflets sometimes divided into 3 parts. Leaflet width variable from 5-15 cm long by 2-6 cm wide. Leaflets may contain small stipule-like tissue at attachment point. Stipule present at leaf base.

Flowers/Fruit
Inflorescence consists of a dense flat topped or gently rounded cyme clustered with small, white, 5-petaled flowers. Fruit is a purple berry. Blooms late April into July, sometimes later; fruits July/August.

Field Characteristics
Distinctive raised lenticels on bark; opposite, compound leaves. Fruit edible upon boiling and used in preserves, wine, and other food; however, the rest of the plant is poisonous if consumed.

Habitat/Range
Common statewide in sunny, wet areas, including freshwater marshes, swamp openings, alluvial forests, and wet pastures; opportunistic in disturbed sites.

Taxonomic Note
Our North Carolina elderberry is *Sambucus nigra* ssp. *canadensis* (synonym: *Sambucus canadensis*).

Symplocos tinctoria — Common Sweetleaf

Leaves pubescent on both sides, to varying degrees

Smooth, brown bark of an older shrub

Flowers open before leaves

Small, oblong fruits, purplish at maturity

Dark green leaves appear clustered at branch ends (shown on forest edge)

SHRUBS

Symplocaceae - Sweetleaf Family

National Wetland Plant List
Mtn/Pdmt: **FAC**   CP: **FAC**

Coefficient of Conservatism
Mtn: **7**   Pdmt: **6**   CP: **6**

Symplocos tinctoria
Common Sweetleaf

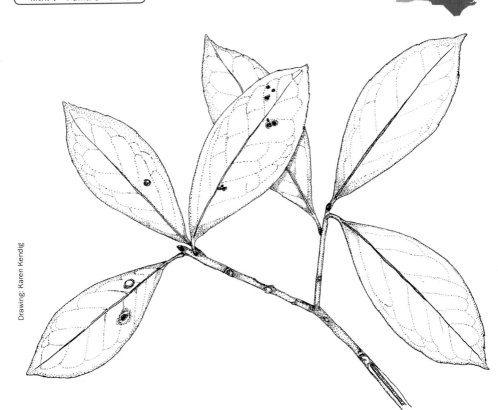

Drawing: Karen Kendig

Habit
Shrub or small tree reaching 8 m in height; tardily deciduous with leaves persisting into winter and a few into spring.

Leaves
Alternate, elliptic or oblanceolate, often with red or purplish blotches. Edges of leaves with minute serrations. Peach-like fuzz on underside of leaves.

Flowers/Fruit
Yellow, fragrant, petal-less flowers with many stamens in spherical clusters close to stem. Fruit is oblong, green drupe, ripening to purple. Blooms March to May; fruits in August/September.

Field Characteristics
As the common name implies, the leaves taste sweet when chewed, making it a favorite of ungulates. Stems have a chambered pith (as illustrated), seen when carefully sliced lengthwise.

Habitat/Range
Damp, yet sandy, soils in mesic woods, ravines, bottomland forests, pine flats, sandy thickets, and pocosin edges, mainly in the Coastal Plain, but still common in the eastern Piedmont. Favors pine-dominated sites.

Similar Species
In some areas, can be confused with *Kalmia latifolia* (mountain laurel) when not in bloom. *K. latifolia* has no spots/glands on leaf undersides, and longer, less elliptical leaves.

Taxonomic Note
This is a monotypic genus - the only genus and species in the sweetleaf family. Also called horse-sugar.

Vaccinium corymbosum — Highbush Blueberry

Deeply veined, light green leaves

White or pink urn-shaped flowers in clusters

Brilliant fall colors typical

Berries blue with glaucous coating

Peeling reddish bark on twigs

SHRUBS

Ericaceae - Heath Family

National Wetland Plant List
Mtn/Pdmt: **FACW** CP: **FACW**

Coefficient of Conservatism
Mtn: **7** Pdmt: **6** CP: **6**

Vaccinium corymbosum
Highbush Blueberry

Habit
Deciduous shrub to 4 m tall with arching green, brown, or red twigs.

Leaves
Alternate, elliptic with entire or finely toothed edges.

Flowers/Fruit
Clusters of small, white or pink, urn-shaped flowers (usually less than 1 cm long). Berries blue with glaucous coating. Flowers in May; fruits mostly in August.

Field Characteristics
Look for blueberries or urn-shaped flowers, characteristic of many members of this family. Brilliant red fall colors are typical. Berries are edible.

Habitat/Range
Swamps, poorly drained wetlands, and bogs; sometimes heath balds or granitic domes, especially near seeps. Mountains and high elevations in western Piedmont.

Similar Species
Gaylussacia spp. (huckleberries) leaves are very similar, but have yellow resinous dots on undersides.

Viburnum nudum — Possumhaw Viburnum

Medium deciduous shrub; floodplains, low woods

Flat-topped clusters of white flowers; glossy leaves

Fruits mature from pink to dark blue

Whitish midvein

Terminal bud distinguishes this from buttonbush, which has similar leaves, but is usually in standing water

SHRUBS

Adoxaceae - Moschatel Family

Viburnum nudum
Possumhaw Viburnum

National Wetland Plant List
Mtn/Pdmt: **OBL**   CP: **FACW**

Coefficient of Conservatism
Mtn: **8**   Pdmt: **7**   CP: **7**

Drawing: Karen Kendig

Habit
Medium sized deciduous shrub up to 5 m tall.

Leaves
Opposite leaves broad, elliptic or obovate with entire or slightly wavy margins. Leaves shiny green, leathery and widest at the middle, 5-10 cm long and 2-6 cm wide.

Flowers/Fruit
Flowers appear in the typical "flat-topped" inflorescence (cyme) and fruits are compressed, blue-black drupes, about 1 cm long. Blooms April/May; fruits August to October.

Field Characteristics
Strictly a wetland species of *Viburnum*. Opposite branching with distinctive appressed or upright buds.

Habitat/Range
Freshwater marshes and swamps, pocosins, wet flats, low woods throughout NC, but especially blackwater floodplains in the Coastal Plain in acidic, stagnant water.

Similar Species
Similar to *Cornus amomum* (silky dogwood) which also has opposite leaves, but more rounded and less shiny leaves. Terminal leaf buds distinguish this from *Cephalanthus occidentalis* (common buttonbush), which has similar leaves.

Viburnum recognitum — Southern Arrowwood

Deeply impressed veins, raised on undersides

Twigs thin, straight, and velvety smooth

Fruit July to September

Smooth leaf petioles have no stipules

Flat-topped clusters of white flowers at branch ends; late March to May

SHRUBS

Adoxaceae - Moschatel Family

National Wetland Plant List
Mtn/Pdmt: **FAC**   CP: **FAC**

Coefficient of Conservatism
Mtn: **7**   Pdmt: **7**   CP: **7**

Viburnum recognitum
Southern Arrowwood

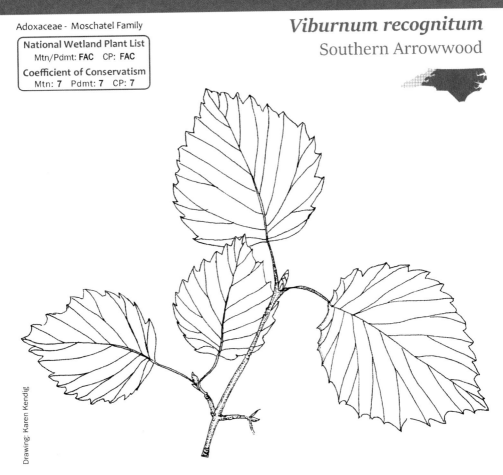

Drawing: Karen Kendig

Habit
Broad-leaved, deciduous shrub to 3 m. Lower stems characteristically straight.

Leaves
Opposite, egg-shaped or almost round, coarse toothed leaves with acute ends. Leaves 5-12 cm long and 4-10 cm wide with serrated edges. Undersides pubescent.

Flowers/Fruit
White flowers form flat-topped clusters at branch ends. Fruit is a medium to very dark blue-gray drupe. Blooms late March to May; fruits July to September.

Field Characteristics
Twigs velvety-hairy and often ridged. The straight wood was formerly used for arrows.

Habitat/Range
Bottomland swamps, shrub wetlands, stream banks, marshes, tidal wetlands. Most abundant in the Piedmont and western half of the Coastal Plain, but in scattered counties in the Mountains and eastern Coastal Plain.

Similar Species
Similar to *Viburnum rafinesquianum* (downy arrowwood) which tends to grow in drier areas, mainly in the Piedmont. Stipules at leaf bases are absent in *V. recognitum* (southern arrowwood) but present in *V. rafinesquianum*, which also has more elongated leaves.

Taxonomic Note
Also called smooth arrowwood or northern arrowwood, this species ranges from eastern Canada to central Georgia and northeastern Alabama. It was recently split out from *Viburnum dentatum*, which does not have smooth (glabrous) petioles. *V. recognitum* was called *V. dentatum* var. *lucidum* in Radford (1968).

FERNS

		National Wetland Plant List Status	
Scientific Name	**Common Name**	**Mtns & Piedmont/Coastal Plain**	**Page**
Athyrium filix-femina	Common Ladyfern	FAC/FAC	166
Onoclea sensibilis	Sensitive Fern	FACW/FACW	168
Osmunda spectabilis	Royal Fern	OBL/OBL	170
Osmundastrum cinnamomeum	Cinnamon Fern	FACW/FACW	172
Woodwardia areolata	Netted Chain Fern	FACW/OBL	174
Woodwardia virginica	Virginia Chain Fern	OBL/OBL	176

Athyrium filix-femina — Common Ladyfern

Basal pinnae point downward; pinnae alternate on stem

Common fern with finely toothed pinnae

Fertile frond with sori

Central stem has two main vascular bundles

Central stem (rachis) often tinged purple; fronds medium light green

FERNS

Dryopteridaceae - Wood Fern Family

National Wetland Plant List
Mtn/Pdmt: **FAC**   CP: **FAC**

Coefficient of Conservatism
Mtn: **6**   Pdmt: **6**   CP: **6**

Athyrium filix-femina
Common Ladyfern

Drawing: Karen Kendig

Habit
Medium light green fine-frond fern to approximately 1 m in height. Grows from rhizomes.

Leaves
Fronds 40-100 cm long and 10-35 cm wide. Because of its finely toothed pinnae, this fern is delicate and lacy in appearance. Pinnae arranged alternately along the stem. Surfaces of fertile and sterile fronds similar in appearance.

Flowers/Fruit
No flowers, but fertile fronds contain crescent-shaped sori on the undersides of the fronds. Sori appear May to September.

Field Characteristics
Attractive, lacy, light green fern. Pinnae and pinnules alternately arranged. Pinnae near the base point downward, less at right angles to the stem than other pinnae. Light green color, and often purple stems, helps distinguish this fern. Stems break easily.

Habitat/Range
Wet woods, streamsides, and swamps throughout NC.

Taxonomic Note
North Carolina's common ladyfern is ssp. *asplenioides*. (Synonym: *Athyrium asplenioides*)

Onoclea sensibilis — Sensitive Fern

Pinnae sometimes mostly untoothed

Fertile fronds appear like clusters of beads

Fronds have overall triangular, or deltoid, shape

Pinnae more or less oppositely arranged

Pinnae sometimes deeply toothed

FERNS

Dryopteridaceae - Dryopteris Fern Family

National Wetland Plant List
Mtn/Pdmt: **FACW**   CP: **FACW**

Coefficient of Conservatism
Mtn: **5**   Pdmt: **5**   CP: **5**

Onoclea sensibilis
Sensitive Fern

Drawing: Karen Kendig

Habit
Low to medium, deciduous fern almost reaching 1 m high, usually smaller. Reproduces by underground rhizomes.

Leaves
Pinnately divided frond. Pinnae more or less positioned oppositely. Margins entire or lobed. Larger plants may have more deeply lobed fronds.

Flowers/Fruit
No flowers, but fertile fronds on separate stalks. Fertile fronds change from green to brown and have a beaded appearance. Fertile fronds appear May through June.

Field Characteristics
Fertile fronds distinctive. Called "sensitive" because of its sensitivity to early frosts. Look for opposite arrangement of pinnae along stem.

Habitat/Range
Marshes, swamps, seeps, moist woodlands, and muddy ditches throughout the state.

Similar Species
Similar to *Woodwardia areolata* (netted chain fern), but pinnae of that fern are alternately arranged and fertile fronds are different. See Common Confusions section, p. 405.

Taxonomic Note
This is the only species within its genus.

Osmunda spectabilis — Royal Fern

Large loose fern with twice divided fronds

Fertile fronds appear March to June

Forms clumps; sometimes in colonies

Pinnule detail

Understory fern of moist woods, swamps, and marshes throughout NC

FERNS

Osmundaceae - Royal Fern Family

National Wetland Plant List
Mtn/Pdmt: **OBL**   CP: **OBL**

Coefficient of Conservatism
Mtn: **7**   Pdmt: **7**   CP: **7**

Osmunda spectabilis
Royal Fern

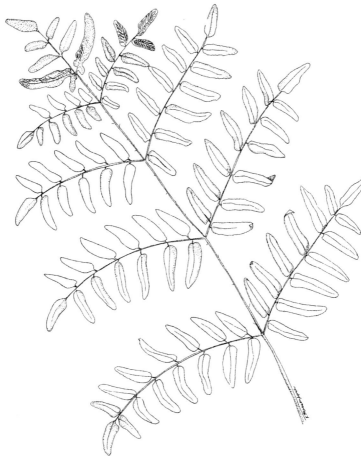

Drawing: Karen Kendig

Habit
Medium to large fern forming clumps. Usually about 1 m tall but may be up to 2 m.

Leaves
Frond is large, 38-75 cm long and 25-50 cm wide. Frond twice pinnately divided. Pinnae usually number 5 to 11, and light green, about 12 cm long and 5 cm wide.

Flowers/Fruit
No flowers, but instead fertile fronds with terminal branch-like panicles. These spore-bearing stalks are light-brown and appear March to June.

Field Characteristics
Large loose fern with twice divided fronds.

Habitat/Range
Moist woods, swamps and marshes throughout NC.

Taxonomic Note
Synonym: *Osmunda regalis* var. *spectabilis*

Osmundastrum cinnamomeum — Cinnamon Fern

Fiddleheads covered with reddish-brown hairs

Cinnamon-colored fertile fronds emerge in spring

Medium to tall fern, up to 1.5 meters high

Hairs at bases of pinnae; no chain-like venation

Large fronds rise singly from small clump at base

FERNS

Osmundaceae - Royal Fern Family

National Wetland Plant List
Mtn/Pdmt: **FACW** CP: **FACW**

Coefficient of Conservatism
Mtn: **7** Pdmt: **6** CP: **7**

Osmundastrum cinnamomeum
Cinnamon Fern

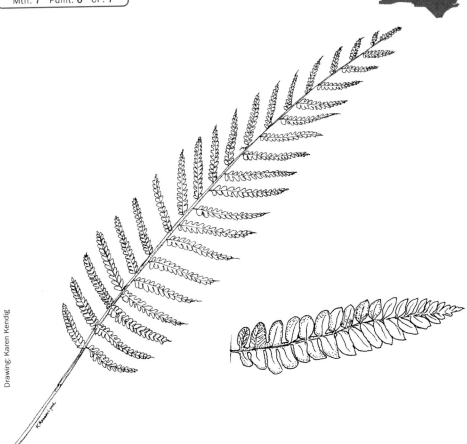

Drawing: Karen Kendig

Habit
Medium to tall fern up to 1.5 m.

Leaves
Fronds rise singly from small clump at base. Blades 35-70 cm long and 13-25 cm wide. Pinnae are cut nearly to stem, mostly alternately arranged with brown, fuzzy "hairs" at bases of pinnae.

Flowers/Fruit
No flowers, but fertile fronds are narrower then infertile fronds and contain furry cinnamon-colored stalks. Fertile fronds appear March to May and soon wither.

Field Characteristics
This is a large fern. When fertile frond is present, it has a distinctive cinnamon color.

Habitat/Range
Swamps, marshes, ditches, streambanks. Common in the Coastal Plain and Mountains, but less frequent in the Piedmont.

Similar Species
Athyrium filix-femina (common ladyfern) has no hairs at base of pinnae; *Woodwardia virginica* (Virginia chain fern) forms clumps and has chain-like venation along midrib of leaflets. See Common Confusions section, p. 406.

Taxonomic Note
Synonym: *Osmunda cinnamomea*

Woodwardia areolata — Netted Chain Fern

Infertile frond

Sometimes in dense colonies in acidic swamps, wet woods

Fertile fronds (last year, current year)

Fertile frond sori like rows of bricks

"Netted" venation and finely serrated edges on pinnae

FERNS

Blechnaceae - Fern Family

Woodwardia areolata
Netted Chain Fern

National Wetland Plant List
Mtn/Pdmt: **FACW**   CP: **OBL**

Coefficient of Conservatism
Mtn: **7**   Pdmt: **6**   CP: **6**

Drawing: Karen Kendig

Habit
Medium fern up to 0.5 m in height, usually smaller. Reproduces by underground rhizomes.

Leaves
Lobed pinnae arranged alternately along stem. Pinnae contain "netted" venation and edges have minute serrations visible when viewed closely.

Flowers/Fruit
No flowers, but fertile fronds on separate stalks which are narrower than sterile fronds. Double rows of sori on fertile frond pinnae resemble rows of bricks. Fertile fronds appear June to September.

Taxonomic Note
Synonym: *Lorinseria areolata* (the only member of *Loniseria* in North Carolina). The genus *Woodwardia* is now considered Eurasian.

Field Characteristics
Look for alternate arrangement of pinnae along stalk and netted venation.

Habitat/Range
Acidic swamps and wet pine woods throughout NC.

Similar Species
Similar in appearance to *Onoclea sensibilis* (sensitive fern), but sensitive fern has opposite pinnae and pinnae edges without serrations, as well as different, bead-like, fertile fronds. See Common Confusions section, p. 405.

Woodwardia virginica — Virginia Chain Fern

Fronds rise singly from ground

Pinnae alternately arranged along stem

Chain-like vein pattern along midrib

Fertile frond sori, some in rows along pinnule stem

Medium-sized fern, to 0.5 m

FERNS

Blechnaceae - Fern Family

National Wetland Plant List
Mtn/Pdmt: **OBL**   CP: **OBL**

Coefficient of Conservatism
Mtn: **n/a**   Pdmt: **7**   CP: **7**

Woodwardia virginica
Virginia Chain Fern

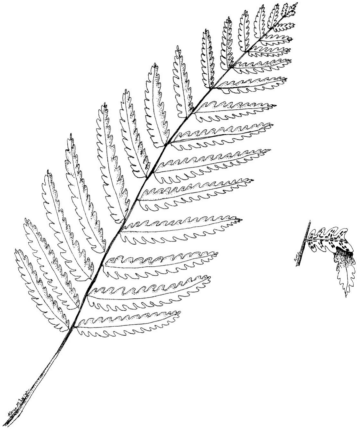

Drawing: Karen Kendig

Habit
Medium fern reaching 0.5 m in height. Reproduces by underground rhizomes.

Leaves
Compound fronds with alternate pinnae. Fronds 30-70 cm long and 15-30 cm wide, rising singly from ground. Leaf base dark, black-brown.

Flowers/Fruit
No flowers, but sori located on undersides of fronds, including rows along pinnule stems. Fertile fronds appear June to September.

Field Characteristics
Look for distinctive chain-like vein pattern on midrib of pinnae.

Habitat/Range
Acidic soils and wet pine flats, mostly in the Coastal Plain. Grows well in sunny locations and responds quickly after fire.

Similar Species
Osmundastrum cinnamomeum (cinnamon fern) appears similar, but fronds of *Woodwardia virginica* (Virginia chain fern) rise singly from the ground, not from a clump. *W. virginica* also has no hairs at bases of pinnae. See Common Confusions section, p. 406.

Taxonomic Note
Synonym: *Anchistea virginica*. The genus *Woodwardia* is now considered Eurasian.

MONOCOT HERBS

Scientific Name	Common Name	National Wetland Plant List Status Mtns & Piedmont/Coastal Plain	Page
Andropogon glomeratus	Bushy Bluestem	FACW/FACW	180
Arisaema triphyllum	Jack-in-the-pulpit	FACW/FACW	182
Aristida stricta	Pineland Threeawn	FAC/FAC	184
Arthraxon hispidus	Small Carpgrass	FAC/FAC	186
Arundinaria gigantea	Giant Cane	FACW/FACW	188
Carex spp.	Common Sedge species	Varies	190
Chasmanthium latifolium	Indian Wood-oats	FACU/FAC	196
Cladium mariscus	Swamp Sawgrass	(n/a)/OBL	198
Commelina virginica	Dayflower	FACW/FACW	200
Cyperus spp.	Common Flatsedge species	Varies	202
Dichanthelium spp.	Rosette Grass	Varies	206
Dichanthelium scoparium	Velvet Panicgrass	FACW/FACW	208
Distichlis spicata	Saltgrass	(n/a)/OBL	210
Dulichium arundinaceum	Three-way Sedge	OBL/OBL	212
Echinochloa crus-galli	Large Barnyard Grass	FAC/FACW	214
Eleocharis obtusa	Blunt Spike-Rush	OBL/OBL	216
Fimbristylis spp.	Common Fimbry species	Varies	218
Hypoxis hirsuta	Common Goldstar	FAC/FACW	220
Iris virginica	Virginia Iris	OBL/OBL	222
Juncus spp.	Common Rush species	Varies	224
Lachnanthes caroliniana	Redroot	OBL/OBL	228
Leersia oryzoides	Rice Cutgrass	OBL/OBL	230
Microstegium vimineum	Japanese Stilt Grass	FAC/FAC	232
Murdannia keisak	Wart-Removing-Herb	OBL/OBL	234
Peltandra virginica	Green Arrow Arum	OBL/OBL	236
Phragmites australis	Common Reed	FACW/FACW	238
Pontederia cordata	Pickerelweed	OBL/OBL	240
Rhynchospora spp.	Common Beaksedge/Beakrush spp.	Varies	242
Saccharum giganteum	Sugarcane Plumegrass	FACW/FACW	246
Sacciolepis striata	American Cupscale	OBL/OBL	248
Sagittaria lancifolia	Bull-tongue Arrowhead	OBL/OBL	250
Sagittaria latifolia	Broadleaf Arrowhead	OBL/OBL	252
Schoenoplectus tabernaemontani	Softstem Bulrush	OBL/OBL	254
Scirpus cyperinus	Woolgrass Bulrush	FACW/OBL	256
Scirpus expansus	Woodland Bulrush	OBL/OBL	258
Scirpus georgianus	Georgia Bulrush	OBL/OBL	260
Sisyrinchium angustifolium	Narrowleaf Blue-eyed Grass	FACW/FACW	262
Sparganium americanum	American Bur-reed	OBL/OBL	264
Spartina alterniflora	Smooth Cordgrass	(n/a)/OBL	266
Spartina cynosuroides	Big Cordgrass	(n/a)/OBL	268
Spartina patens	Saltmeadow Cordgrass	(n/a)/FACW	270
Typha angustifolia	Narrowleaf Cattail	(n/a)/OBL	272
Typha latifolia	Broadleaf Cattail	OBL/OBL	274
Xyris spp.	Yellow-eyed Grass	Most OBL/Most OBL	276
Zephyranthes atamasca	Atamasco Lily	FACW/FACW	278

Andropogon glomeratus — Bushy Bluestem

Nearly mature inflorescences

Mature inflorescences fluffy and full

A common fall sight along roadside swales and ditches

Last year's inflorescence

Leaves have bluish or whitish coating (glaucous)

MONOCOT HERBS

Poaceae - Grass Family

Andropogon glomeratus
Bushy Bluestem

National Wetland Plant List
Mtn/Pdmt: **FACW**   CP: **FACW**

Coefficient of Conservatism
Mtn: **5**   Pdmt: **3**   CP: **3**

Drawing: Mark Mohlenbrock, USDA Wetland Flora Field Office Guide

Habit
Medium to tall perennial grass, 1 to 2 m high, growing in dense clumps.

Leaves
Leaves up to 30 cm long, with broad overlapping sheaths. Chalky coating makes the leaves appear bluish-green.

Flowers/Fruit
Prominent, feathery inflorescence, branched but appearing dense, and usually 2-5 cm long. Flowering and fruiting late August-October.

Field Characteristics
Inflorescence has paired spikelets; one of each pair has a long hairy bristle extending from its base.

Habitat/Range
Roadside swales, fields, open woods, savannas, and wetlands. Less common in the mountains, but widespread across the Piedmont and Coastal Plain.

Similar Species
An uncommon coastal species, *Andropogon glaucopsis* (purple bluestem), is now recognized as distinct. It has a less dense inflorescence with a purple hue in fall and long hairs on stem branches below raceme bracts.

Arisaema triphyllum — Jack-in-the-pulpit

Often grows in small colonies

Emerging leaf

Spathe can be purple or green and white striped

Green fruits mature to bright red

Compound leaves with three leaflets, smooth margins; main veins do not extend to edges of leaflets

MONOCOT HERBS

Araceae - Arum Family

National Wetland Plant List
Mtn/Pdmt: **FACW**   CP: **FACW**

Coefficient of Conservatism
Mtn: 6   Pdmt: 6   CP: 6

Arisaema triphyllum
Jack-in-the-pulpit

Highly Toxic Plant

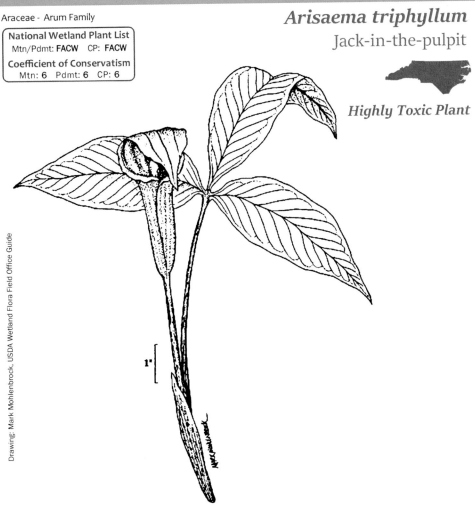

Habit
Small to medium herb, characterized by a pair of compound leaves with three leaflets and a flowering spathe with hood. To 60 cm tall.

Leaves
Three leaflets, margins entire, palmately compound. Leaves whitish beneath at maturity.

Flowers/Fruit
Flowering body is in the form of a cup-like structure (spathe) with an overhanging hood and a stiff projection (spadix) in the center. Plants produce individual male or female flowers, but can switch from year to year. Bright red berries appear in late summer and fall. Blooms in March/April; fruits soon after.

Field Characteristics
Look for three leaflets in compound leaves, with smooth margins and rising singly from the ground. Leaves and fruits are highly toxic if ingested.

Habitat/Range
Swamps and bottomlands; common across the state.

Similar Species
Leaflets in sets of three may be confused with *Toxicodendron radicans* (eastern poison ivy), but *Arisaema* (Jack-in-the-pulpit) never has lobed leaves like *T. radicans* does, and in *T. radicans*, leaf veins extend to the edges of the leaves. See Common Confusions section, p. 410.

Taxonomic Note
USDA lists several subspecies that Weakley splits out into separate species, varying by size and hood/spathe characteristics.

Aristida stricta   Pineland Threeawn

New inflorescences

Older inflorescence

Restricted to wet pine flats and related habitat

Low intensity burns do not kill the plant

Clumping perennial grass, 0.5 to 1 m

MONOCOT HERBS

Poaceae - Grass Family

National Wetland Plant List
Mtn/Pdmt: **FAC**   CP: **FAC**
Coefficient of Conservatism
Mtn: **n/a**   Pdmt: **8**   CP: **8**

Aristida stricta
Pineland Threeawn

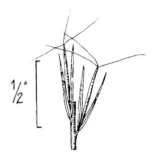

Habit
Medium height perennial grass forming graceful tufts or clumps, 0.5 to 1 m tall.

Leaves
Long, narrow leaf blades rolled inward and mostly originating from the base. Dense leaf hairs evident without unrolling the blade.

Flowers/Fruit
Loosely arranged spikelets in a slender panicle, up to 30 cm long. Flowers of this genus have 3 distinctive awns or bristles about 1 cm long. Flowering time depends on burning; flowers July to October if burned earlier in the year.

Field Characteristics
Easy to distinguish when in flower since it is restricted to wet pine flats and related habitat. This species only flowers after its habitat has been burned.

Habitat/Range
Wiregrass tolerates a wide range of moisture including well drained hills, wet pine savannas, and flatwoods, especially where periodically burned. Mainly found in the Coastal Plain.

Similar Species
Can be confused with *Sporobolus pinetorum* (Carolina dropseed), but that uncommon species is more green.

Arthraxon hispidus — Small Carpgrass

Fan-shaped inflorescence with thin spikelets

Flowers and fruits in fall

Hairs on leaf sheaths and leaf edges

Long, thin, creeping stems

Grows in colonies

MONOCOT HERBS

Poaceae - Grass Family

National Wetland Plant List
Mtn/Pdmt: **FAC**   CP: **FAC**

Coefficient of Conservatism
Mtn: **0**   Pdmt: **0**   CP: **0**

Arthraxon hispidus
Small Carpgrass

Non-native

Habit
Low annual with thin, creeping, branching stems, 20-100 cm long. Stems root at the nodes.

Leaves
Leaf blades ovate or lance-shaped, 2-5 cm long and 0.5-1.5 cm wide.

Flowers/Fruit
Fan-shaped inflorescence formed by stalks of thin spikelets. Individual spikelets 4-5 mm long, sometimes pink. Flowering and fruiting September to November.

Field Characteristics
The creeping nature of this plant along with the ovate leaves are distinctive features.

Habitat/Range
This introduced species inhabits waste areas, rapidly spreading throughout ditches and wet areas.

Similar Species
Similar to another non-native, *Microstegium vimineum* (Japanese stilt grass), but *Arthraxon hispidus* (small carpgrass) has wider leaves with hairy sheaths clasping the stems. *M. vimineum* also has leaves with shiny midribs.

Arundinaria gigantea — Giant Cane

Long hairs where leaves meet stems

Can be found in extensive colonies

Spreads primarily through rhizomes

Narrow, unbranched stems, usually 1 to 3 m

Seldom seen in flower or fruiting

MONOCOT HERBS

Poaceae - Grass Family

> **National Wetland Plant List**
> Mtn/Pdmt: **FACW** CP: **FACW**
>
> **Coefficient of Conservatism**
> Mtn: **6** Pdmt: **6** CP: **5**

Arundinaria gigantea
Giant Cane

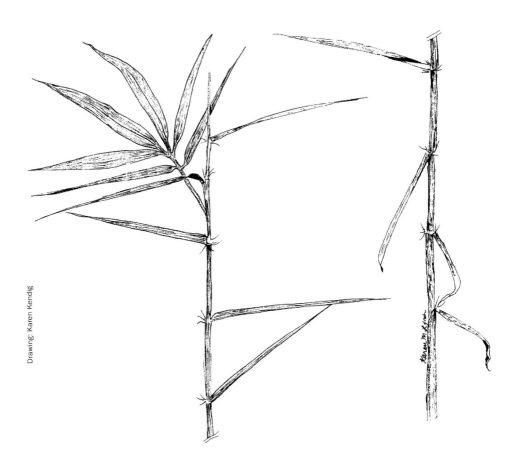

Drawing: Karen Kendig

Habit
Tall, thick, woody herb, usually 1-3 m but may grow much taller (to 7 m). Forms extensive colonies since it reproduces mainly by rhizomes.

Leaves
Alternate leaves with flat acuminate blades. Leaf blades 15-25 cm long and 2-4 cm wide with rounded bases.

Flowers/Fruit
Solitary grass-like spikelets with brown grains. Spreads primarily through rhizomes; fruits mainly just before death, which can be after 4 to 5 decades.

Field Characteristics
Culms were historically used for fishing poles, pipe stems, baskets, and mats.

Habitat/Range
Low woods, rich soils along brownwater rivers, bottomlands throughout NC.

Similar Species
Resembles the non-native Chinese bamboo, which grows much taller and in very dense colonies.

Taxonomic Note
Both *Arundinaria gigantea* ssp. *tecta* and *A. gigantea* ssp. *gigantea* are found in North Carolina.

Common *Carex* species

Cyperaceae - Sedge Family

Carex spp.
Sedge

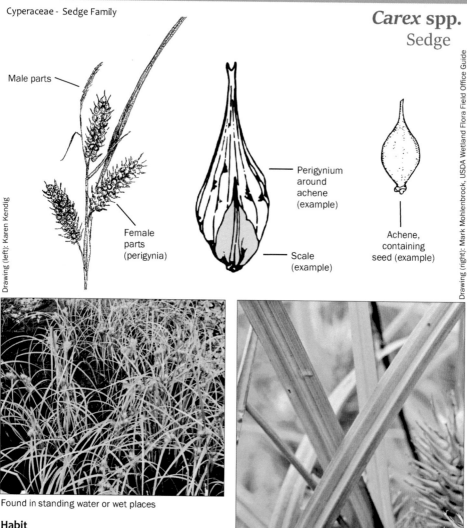

Found in standing water or wet places

Indented (keeled) leaves

Habit
Medium to tall grass-like sedge, 15 to 120 cm tall. Sedges may spread by rhizomes but most commonly reproduce by seed and form clumps. Stems solid and triangular.

Leaves
Thin, linear leaf blades with distinct linear indent, or keel, at midrib. Leaves typically in multiple directions, often overtopping inflorescences.

Flowers/Fruit
Inflorescence contains male and female flowers crowded in separate cylindrical clusters on the same plant. All *Carex* species share the feature of seeds (achenes) being completely encompassed by an outer covering (perigynium). Fruiting in late spring and early summer.

Field Characteristics
Leaf blades "v" shaped with distinctive keels. Flowering stems triangular. Identification to species requires mature fruits and seeds (achenes).

Habitat/Range
The genus *Carex* contains the most members of the sedge family and is the largest single genus in North Carolina. Found in standing water or in wet soils, in shade and sun.

MONOCOT HERBS

Carex abscondita - Thicket Sedge

Mtn/Pdmt-FAC/CP-FACW   C values: 7 - 8 - 8

A low, flattened, spreading woodland plant with wide, leaves and long extended flowering stems with clusters of plump perigynia.

Carex atlantica - Prickly Bog Sedge

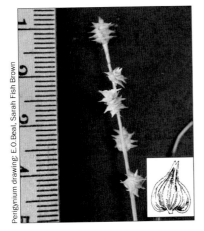

Mtn/Pdmt-FACW/CP-FACW   C values: 7 - 7 - 7

A medium-sized, fine leaved sedge with groups of spiny, toothed perigynia on long flowering stems. Swamps, pocosins, seepage bogs.

Carex blanda - Common Woodland Sedge

Mtn/Pdmt-FAC/CP-FAC   C values: 4 - 4 - 4

A small, clumping, woodland sedge with narrow leaves and long flowering stems with crowded clumps of perigynia, each wider past the middle.

Common *Carex* species

Carex caroliniana - Carolina Sedge

Mtn/Pdmt-FACW/CP-FACW   C values: 5 - 5 - 5

A medium to small spreading sedge, with wide leaves and densely packed, fat perigynia in seedheads. Ditches, damp meadows, bottomlands; sometimes drier places.

Carex crinita - Fringed Sedge

Mtn/Pdmt-OBL/CP-FACW   C values: 6 - 6 - 6

A large, upright, clumping shade and water loving sedge. Each perigynium has a long scale attached, hence the "fringed" appearance. The similar *Carex gynandra* (nodding sedge) occurs in the mountains.

Carex debilis - White-Edge Sedge

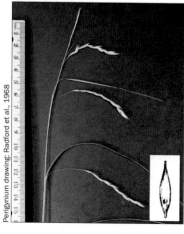

Mtn/Pdmt-FAC/CP-FACW   C values: 6 - 6 - 6

A medium-sized clumping sedge of moist woods with drooping seedheads. The uppermost part of the inflorescence has only male parts, and lower ones are female, producing seeds.

MONOCOT HERBS

Carex glaucescens - Southern Waxy Sedge

Mtn/Pdmt-OBL/CP-OBL   C values: absent - 7 - 7

A large, blue-green leaved sedge with drooping, spiky, whitish seedheads. Sunny or partly sunny marshes, acidic streamheads, wet ditches.

Carex intumescens - Greater Bladder Sedge

Mtn/Pdmt-FACW/CP-FACW   C values: 6 - 6 - 6

A medium to small sparse sedge. Seedheads with few large perigynia. Similar to the less common *Carex grayi* (Gray's sedge), which has downward-pointing perigynia, often finely haired. Shaded freshwater wetlands of many types.

Carex laevivaginata - Smooth-sheathed Sedge

Mtn/Pdmt-OBL/CP-OBL   C values: 7 - 7 - 7

A medium-sized clumping sedge with spiky seedheads and wide, easily compressed stems. Look for unwrinkled leaf sheath around lower part of flowering stem. Partly sunny spots in swamps, marshes.

Common *Carex* species

Carex lupulina - Hop Sedge

Mtn/Pdmt-OBL/CP-OBL   C values: 5 - 5 - 5

A large, upright sedge with very large, spiny seedheads. Perigynia oriented in upward direction. Sunny wet marshes, bottomlands.

Carex lurida - Shallow Sedge

Mtn/Pdmt-OBL/CP-OBL   C values: 5 - 4 - 4

A large, bright green sedge, with deeply grooved leaves and large, shiny, spiny seedheads, sometimes quite elongated. Perigynia oriented more at right angle to stem. Sunny wet places.

Carex oxylepis - Sharpscale Sedge

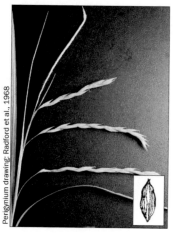

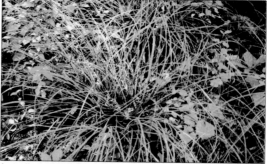

Mtn/Pdmt-FACW/CP-FACW   C values: 7 - 7 - 7

A medium-sized, fine leaved, woodland sedge. Leaves covered with soft short hairs. Uppermost seedhead has both male and female parts.

MONOCOT HERBS

Carex tribuloides - Blunt Brown Sedge

Mtn/Pdmt-FACW/CP-FACW C values: 4 - 4 - 5

A medium to large, grass-like, non-clumping sedge with egg-shaped seedheads; perigynia winged, but not to base (similar to *Carex scoparia* [broom sedge] which is winged to base). Brownwater floodplains and bottomlands.

Carex vulpinoidea - Common Fox Sedge

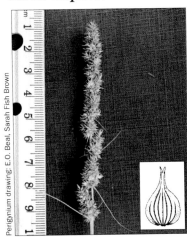

Mtn/Pdmt-OBL/CP-FACW   C values: 3 - 3 - 3

A medium to large clumping sedge with leaves longer than flowering stems. Many bristles extend outward from dense inflorescences. Individual perigynia quite flattened. Open, sunny, freshwater wetlands.

Chasmanthium latifolium — Indian Wood-oats

Flowers and fruits June to October

Often in loose colonies

Loose, long hairs where leaves clasp stem

Flattened seedheads droop from long stalks

Seeds turn pale brown in fall

MONOCOT HERBS

Poaceae - Grass Family

National Wetland Plant List
Mtn/Pdmt: **FACU**   CP: **FAC**

Coefficient of Conservatism
Mtn: **6**   Pdmt: **6**   CP: **6**

Chasmanthium latifolium
Indian Wood-oats

Drawing: Karen Kendig

Habit
Tall, colonial grass, spreading by rhizomes. Plant unbranched with stems 0.6 to 1.5 m tall.

Leaves
Grass-like leaf blades, 2 cm wide. Blades smooth with rough edges. Leaf bases have long, loose hairs.

Flowers/Fruit
Flowering portion with drooping branches containing spikelets, which resemble sea oats. Flowers and fruits June to October.

Field Characteristics
Look for long loose hairs where leaves clasp stem; smooth leaves with rough edges. Fruits distinctive.

Habitat/Range
Stream and river banks, low woods and shaded slopes throughout the state, but more numerous in the Piedmont.

Similar Species
Similar in appearance to the common coastal *Uniola paniculate* (sea oats), but grows in freshwater wetlands.

Cladium mariscus — Swamp Sawgrass

Tall perennial grass to 3 m

Inflorescences are clusters of rounded spikelets

Often in extensive stands

Light and dry in winter

Saw-like teeth on both edges of blades

MONOCOT HERBS

Cyperaceae - Sedge Family

Cladium mariscus
Swamp Sawgrass

National Wetland Plant List
Mtn/Pdmt: n/a   CP: **OBL**

Coefficient of Conservatism
Mtn: **n/a**   Pdmt: **n/a**   CP: **8**

Drawing: Karen Kendig

Habit
Tall, coarse perennial sedge up to 3 m growing in extensive stands from stout rhizomes. Stem is slightly triangular.

Leaves
Linear tapering leaves, about a meter long and one centimeter wide with spiny, saw-toothed margins. Leaves are stiff, folded at the midrib and becoming triangular at the tip.

Flowers/Fruit
Long (0.5 m) inflorescence is formed by clusters of spikelets occurring at the end of drooping branches. Flowering and fruiting July to September.

Field Characteristics
Leaf margins feel unmistakably like a saw; walking through a sawgrass marsh can be a painful experience.

Habitat/Range
Brackish marshes, ditches and shores in the outer Coastal Plain. May form dense monotypic stands in brackish waters. May occur either in standing water or on less wet ground in the Coastal Plain.

Taxonomic Note
Synonym for *Cladium mariscus* spp. *jamaicense* is *Cladium jamaicense*.

Commelina virginica — Dayflower

Leaf sheaths with coarse reddish hairs

Solitary blue flowers, one petal smaller than others

Often in dense colonies

Flowers and fruits July to October

Large, flattened seed-containing sheaths

MONOCOT HERBS

Commelinaceae - Dayflower Family

National Wetland Plant List
Mtn/Pdmt: **FACW**   CP: **FACW**
Coefficient of Conservatism
Mtn: **6**   Pdmt: **5**   CP: **5**

Commelina virginica
Dayflower

Drawing: Kristie Gianopulos

Habit
Low to medium height herb with wide, simple leaves, succulent stems, and two or three-petaled blue flowers. Often recumbent.

Leaves
Alternate, lance-shaped, simple, with entire margins and parallel veins. Leaf sheaths extend below leaf base onto stem; sheaths with short, coarse, reddish hairs.

Flowers/Fruit
Solitary blue flowers with two or three frilly petals, one smaller than the other two. Fruits fairly large flattened sheaths with 2-3 sides, containing brown or reddish seeds. Blooms and fruits July to October.

Field Characteristics
Two-petaled flowers are distinctive; look for leaf sheaths extending below leaf bases.

Habitat/Range
Bottomlands, wet forests, and forest edges.

Similar Species
Six species of *Commelina* occur in North Carolina; two are non-native. The most widespread species are *Commelina communis* (Asiatic dayflower; non-native), *C. diffusa* (climbing dayflower), *C. erecta* (whitemouth dayflower), and *C. virginica* (Virginia dayflower), all differing slightly in flower details and size. In *C. communis* and *C. erecta* the third, lower petal is white.

Common *Cyperus* species

Cyperaceae - Sedge Family

Cyperus spp.
Flatsedge
Some non-native

Terminal inflorescences with flattened spikelets

Usually 4 (+) leaves extend from base of inflorescence

Found in standing water or wet places

Sedges usually have triangular flowering stems

Habit
Clumping, perennial, grass-like plant (sedge), with triangular stems, sometimes forming loose colonies.

Leaves
Basal linear leaves. Usually 4 or more leaves extending out from base of inflorescence.

Flowers/Fruit
Inflorescences located at ends of stems, with flattened spikelets usually in pairs or groups. Central group of spikelets with very short connecting stalk or none at all. Seeds in a folded scale, unlike *Carex*.

Field Characteristics
Cyperus species vary greatly in size. Identification to species requires examination of mature seeds and fruits.

Habitat/Range
Roadside ditches, low fields, marshes; generally restricted to wetter conditions.

MONOCOT HERBS

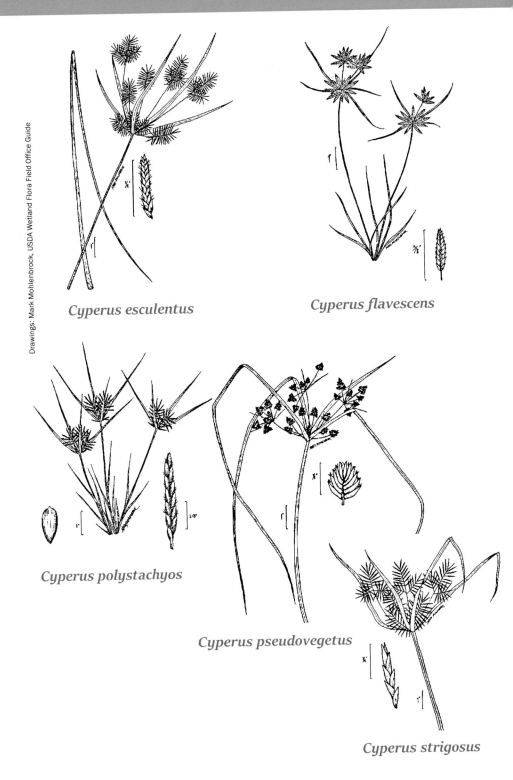

Cyperus esculentus

Cyperus flavescens

Cyperus polystachyos

Cyperus pseudovegetus

Cyperus strigosus

Common *Cyperus* species

Cyperus esculentus - Chufa; Yellow Nutsedge

Non-native
Mtn/Pdmt-FACW/CP-FAC   C values: 1-1-1

A large, *non-native* sedge, with airy seedheads, composed of long, thin spikelets. Often planted; can grow in fields, disturbed places, also in wetlands. Underground nutlets form along roots.

Cyperus flavescens - Yellow Flatsedge

Mtn/Pdmt-OBL/CP-OBL   C values: 3-4-4

A short, weedy sedge with long flattened spikelets; sunny low fields, ditches, marshes, mainly in the Mountains and Piedmont.

MONOCOT HERBS

Cyperus polystachyos - Manyspike Flatsedge

Mtn/Pdmt-FACW/CP-FAC   C values: 3-4-4

Densely packed seedheads with long spikelets. Sunny shores, ditches, brackish marshes.

Cyperus pseudovegetus - Marsh Flatsedge

Mtn/Pdmt-FACW/CP-FAC   C values: 3-3-3

Spikelets flattened and rounded; densely packed. Sunny marshes, ditches, depressions.

Cyperus strigosus - Straw-color Flatsedge

Mtn/Pdmt-FACW/CP-FACW   C values: 2-2-3

A large sedge, with prolific oblong-shaped seedheads. Sunny shores, marshes, ditches, wet flatwoods.

Dichanthelium spp.   Rosette Grass

Clumping perennial

Soft, lance-shaped leaves and hairy stems

Basal rosette

Egg-shaped fruit in loose panicles with leaf at base

Immature inflorescence; flowering and fruiting May to October

MONOCOT HERBS

Poaceae - Grass Family

Dichanthelium spp.
Rosette Grass

National Wetland Plant List
Mtn/Pdmt: **Varies** CP: **Varies**
Coefficient of Conservatism
Varies

Drawing: Karen Kendig

Habit
Clumping, perennial grass with soft, lance-shaped leaves and hairy stems.

Leaves
Lance-shaped leaves at plant base typically form a basal rosette of short blades.

Flowers/Fruit
Egg-shaped fruit in a loose panicle born on hairy stems with a leaf attached at the base of the panicle. Fruits are seeds enclosed in specialized structures. Flowering and fruiting May to October.

Field Characteristics
Look for rosette clumps of lance-shaped leaves, low to the ground and fuzzy stems with loose seed clusters.

Habitat/Range
Many species of *Dichanthelium* (rosette grass) tolerate drier conditions, but this genus can be found throughout the state in moist woods, wet fields, moist sandy soil, ditches, and edges of pocosins and pine wetlands.

Taxonomic Note
Some taxonomic confusion exists among *Dichanthelium* species, which are notoriously difficult to tell apart, but Weakley (2020) has a good key.

Dichanthelium scoparium — Velvet Panicgrass

Erect branching form

Velvet hairy on stems and both sides of leaves

Distinctive sticky band below nodes, without hairs

Typical loose branching panicle

Sometimes forms loose colonies

MONOCOT HERBS

Poaceae - Grass Family

Dichanthelium scoparium
Velvet Panicgrass

National Wetland Plant List
Mtn/Pdmt: **FACW**   CP: **FACW**

Coefficient of Conservatism
Mtn: 5   Pdmt: 5   CP: 5

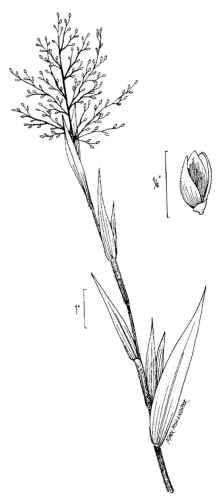

Habit
A late-branching, leafy plant, somewhat erect, to 1 or 1.25 m tall.

Leaves
Leaves and stems feel like velvet; blades 20 cm long and 10 to 20 mm wide.

Flowers/Fruit
Typical of *Dichanthelium* species, loose branching panicle at branch tips. Flowering and fruiting May to October.

Field Characteristics
Look for distinctive clear or sticky ("viscid") band below nodes on stems; leaves and stems velvet hairy.

Habitat/Range
Ditches, low woods, marshes, wet savannas or pastures, openings in swamp forests, wet disturbed areas; mainly Coastal Plain and eastern Piedmont. Present, but uncommon to rare in the western Piedmont and Mountains.

Similar Species
Distinguishable from other *Dichanthelium* species by its non-hairy, sticky band below the nodes. Refer to Weakley's key for identifying other *Dichanthelium* species.

Distichlis spicata — Saltgrass

Leaves thin, arranged in one plane along stiff stem

Male and female flowers on separate plants

Forms dense colonies, spreading by rhizomes

Flattened seedheads

Edges of leaves rolled inward; can tolerate high salinities

MONOCOT HERBS

Poaceae - Grass Family

Distichlis spicata
Saltgrass

National Wetland Plant List
Mtn/Pdmt: n/a   CP: OBL

Coefficient of Conservatism
Mtn: n/a   Pdmt: n/a   CP: 8

Drawing: Sara Fish Brown/E.O. Beal

Habit
Short wiry grass 10-40 cm tall. Forms dense colonies spreading by stout rhizomes. Runners on the ground sometimes evident.

Leaves
Numerous thin, linear leaves angle at 45 degrees to the stem. Leaves distinctly 2-ranked or arranged in one plane on opposite sides of the stiff, hollow stem. Leaves contain overlapping sheaths and edges of leaves are rolled inward.

Flowers/Fruit
Terminal inflorescence is a light green panicle, 1 to 6 cm long. Male and female flowers occur on separate plants with small spikelets of 5-9 flowers. Flowers and fruits June-October.

Field Characteristics
Angles of leaves are distinctive forming a "v" against the stem. *Distichlis spicata* (saltgrass) is commonly found with *Spartina patens* (saltmeadow cordgrass), but usually "hidden" among the taller grasses with which it grows.

Habitat/Range
Salt marshes or brackish marshes, along seashores, forming dense colonies. Can tolerate salinities exceeding that of full strength seawater.

Taxonomic Note
Saltgrass is named from the Greek word, "distichos", meaning leaves are 2-ranked (arranged in one plane on opposite sides of the stem).

Dulichium arundinaceum — Three-way Sedge

Leaves branch off stem in three directions

Flowers with spikelets on upper part of stem

Found in colonies, spreading by rhizomes

Unbranched flowering stalk

Narrow, grooved, linear leaves

MONOCOT HERBS

Cyperaceae - Sedge Family

National Wetland Plant List
Mtn/Pdmt: **OBL**   CP: **OBL**

Coefficient of Conservatism
Mtn: **7**   Pdmt: **7**   CP: **6**

Dulichium arundinaceum
Three-way Sedge

Drawing: Courtesy of the Flora of North America Association, Yevonn Wilson-Ramsey; habit - Sara Fish Brown/E.O. Beal

Habit
Medium, upright leafy sedge, to 60 cm tall.

Leaves
Narrow, grooved linear leaves, pointed in three main directions from stem.

Flowers/Fruit
Flower on upper part of stem, with slender, brownish spikelets emerging from leaf bases. Flowers and fruits July-October.

Field Characteristics
This sedge is recognizable by its leaves pointed in three distinct directions, when viewed from above. Spreads by rhizomes; usually in colonies.

Habitat/Range
Swamp forests, marshes, beaver ponds, mountain bogs, wet ditches, mainly in the Coastal Plain; uncommon elsewhere across the state.

Taxonomic Note
This is the only species in the genus *Dulichium*.

Echinochloa crus-galli Large Barnyard Grass

Terminal panicle inflorescence, 10 to 25 cm long

Inflorescences can have reddish tint

Spined seed case (left) with rounded seed inside (right)

Loose habit

White line on leaf blades

MONOCOT HERBS

Poaceae - Grass Family

National Wetland Plant List
Mtn/Pdmt: **FAC** CP: **FACW**

Coefficient of Conservatism
Mtn: **0** Pdmt: **0** CP: **0**

Echinochloa crus-galli
Large Barnyard Grass

Non-native

Habit
Medium, upright, annual grass, with stout stems, up to over 1 m tall. Frequently branching at the base.

Leaves
Long, tapering leaves, up to 0.5 m long and 1-2 cm wide.

Flowers/Fruit
Compact terminal panicle, 10 to 25 cm long. Inflorescence appears purplish with bristles and erect or nodding. Spikelets densely concentrated on one side of flowering branches. Seed cases can have bristles of widely varying lengths. Seeds are rounded with tiny tufts of hairs at tips. Flowering and fruiting July-November.

Field Characteristics
Barnyard grass is commonly planted on wildlife refuges as the seeds are utilized by waterfowl and other birds.

Habitat/Range
Fresh marshes, slightly brackish marshes, swamps, moist open areas and waste places.

Similar Species
Easily confused with the equally common **native** *Echinochloa muricata* (rough barnyard grass), which has more elongated seeds, in seed cases with larger spines that have swollen bases (requiring magnification to see). Excellent photographic comparisons can be found on the website Minnesotawildflowers.info.

Taxonomic Note
Synonym: *Echinochloa crusgalli*

Eleocharis obtusa — Blunt Spike-Rush

Clumping rush with thin, erect stems

Tiny cone-like seedheads, not much wider than stem

No petals on seedheads; orange-brown overlapping scales

Stems emerge from base, to 30 cm

Prefers sunny wet spots, including marsh edges, ditches

MONOCOT HERBS

Cyperaceae - Sedge Family

Eleocharis obtusa
Blunt Spike-Rush

National Wetland Plant List
Mtn/Pdmt: **OBL** CP: **OBL**

Coefficient of Conservatism
Mtn: **3** Pdmt: **4** CP: **3**

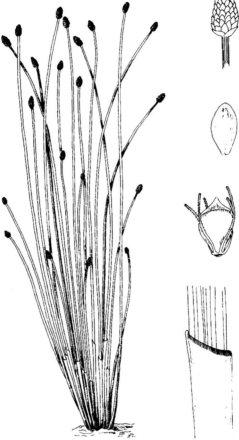

Drawing: Sara Fish Brown/E.O. Beal

Habit
Clump-forming perennial, composed of thin, straight, erect stems emerging from the base, up to 30 cm.

Leaves
Leaves reduced to sheaths on stems; most of above-ground vegetation of the plant is fine, smooth stems.

Flowers/Fruit
Oval or egg-shaped seedhead born on a single spike with no leaves, leaflets, or petals. Seedhead has orange-brown overlapping scales like an immature pine cone, usually spirally arranged. Flower and fruit June to October.

Field Characteristics
Smooth, fine stems with no apparent leaves and single, small, cone-like seedheads help distinguish *Eleocharis* spp. from other grass-like plants. Identification to species generally requires examination of mature seeds and fruit capsules.

Habitat/Range
Edges of marshes, swamps, ditches, often in shallow water in sandy or peaty soil.

Similar Species
Xyris spp. (yellow-eyed grass) also have single cone-like seedheads at the ends of stems, but erect flat leaf blades grow with stems. *Xyris* spp. also have yellow petals on the flowering heads.

Fimbristylis spp. — Fimbry

Fimbristylis autumnalis, a small statewide species

Fimbristylis castanea, a large coastal species

Clumping plant, seed-bearing stems exceed leaves in length

Fimbristylis autumnalis

Fimbristylis castanea

MONOCOT HERBS

Cyperaceae - Sedge Family

Fimbristylis spp.
Fimbry

National Wetland Plant List
Mtn/Pdmt: **Varies**   CP: **Varies**

Coefficient of Conservatism
Mtn: **Varies**   Pdmt: **Varies**   CP: **Varies**

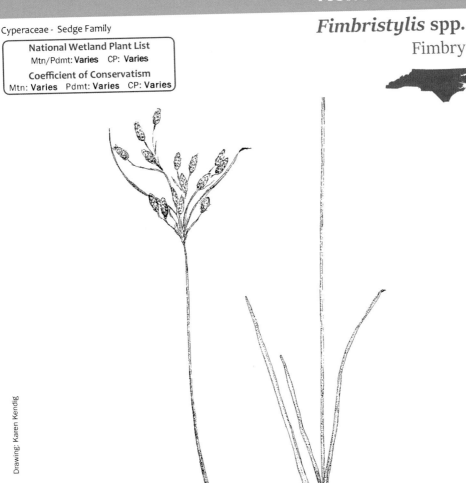

Drawing: Karen Kendig

Habit
Clumping annual, with stems holding loose clusters of seeds extending out beyond the leaves; up to 50 cm tall.

Leaves
Blade leaves relatively short, with much longer flowering stems emerging beyond blades. Flowering stems are flattened.

Flowers/Fruit
Seedheads, without bristles, very small and cone-shaped, on numerous airy branching panicles. Flowering and fruiting June to October.

Field Characteristics
With this clumping grass, look for loose branching clusters of small seeds at tips of flattened stems. *Fimbristylis castanea* (marsh fimbry) is a large common coastal species and *F. autumnalis* (slender fimbry) is a small common species statewide.

Habitat/Range
Found in moist to wet soil of any type; meadows, ditches, low disturbed areas. *Fimbristylis castanea* (marsh fimbry) is found in brackish and fresh-tidal marshes of the Coastal Plain. *F. castanea* (C value: 7)(NWPL: OBL); *F. autumnalis* (slender fimbry) (C values: 4-3-3)(NWPL: Mtn/Pdmnt - FACW; CP - OBL).

Similar Species
Other *Fimbristylis* species occurring in North Carolina are uncommon to very rare.

Hypoxis hirsuta — Common Goldstar

Found in wide variety of dry and moist habitats

Flowers March to June

Leaves grass-like; hairy

Low-growing plant

Yellow, six-petaled flowers with six anthers

MONOCOT HERBS

Hypoxidaceae - Star-grass Family

National Wetland Plant List
Mtn/Pdmt: **FAC**   CP: **FACW**

Coefficient of Conservatism
Mtn: **6**   Pdmt: **5**   CP: **5**

Hypoxis hirsuta
Common Goldstar

Drawing: Mark Mohlenbrock, USDA Wetland Flora Field Office Guide

Habit
Low-growing, grass-like plant with yellow flowers.

Leaves
Leaves grass-like, hairy, growing in a cluster. Usually 22 to 30 cm long and less than 1 cm wide.

Flowers/Fruit
Flowering stem usually shorter than leaves, 15 to 20 cm tall. Yellow 6-petaled flowers with 6 anthers. Blooms March to June; fruits May/June.

Field Characteristics
This is the most abundant star-grass in the state.

Habitat/Range
Found in a wide variety of dry and moist habitats, usually in part sun, along forest edges, meadows, sometimes in bottomland forests and savannas.

Common *Juncus* species

Juncaceae - Rush Family

Juncus spp.
Rush

J. coriaceus (left); *J. effusus* (right). Some species have terminal inflorescences (unlike those shown)

Some *Juncus* species have septate leaves (cross partitions)

Needle-like stems usually cylindrical in cross section; sometimes flattened

Habit
Annual or perennial (mostly perennial and reproduce by rhizomes) grass-like rushes of wet areas, forming dense clumps. Upright, cylindrical, hollow, smooth stems usually pale or bright green and needle-like.

Leaves
Needle-like stems usually cylindrical but in some species flattened. In some species, leaf blades absent, but brownish-red leaf sheaths that open vertically are found near stem bases Flowering stalks cylindrical and smooth.

Flowers/Fruit
Inflorescence is panicle of numerous small flowers on stalks of varying lengths. In some species, a bract extends above flowering stem. In others, there is no bract and inflorescence appears terminal. All *Juncus* have 6 sepal-like structures surrounding each fruit.

Field Characteristics
Juncus species are grouped into two major groups, based on whether leaves have internal divisions (septate leaves). Identification to species requires examination of mature seeds and fruits.

Habitat/Range
Forms large clumps, usually in sunny or partly sunny places, in marshes, along pond/lake edges and in wet fields.

MONOCOT HERBS

Juncus acuminatus - Knotty-Leaf Rush

Mtn/Pdmt-OBL/CP-OBL   C values: 4 - 4 - 4

Stiff, globular seedheads on a open, terminal inflorescences; septate leaves. Wide variety of open, wet areas.

Juncus biflorus - Bog Rush

Mtn/Pdmt-FACW/CP-FACW   C values: 5 - 5 - 5

Flattened, grass-like, non-septate leaves; terminal inflorescence. Grows to 0.5 to 0.8m. Very similar to the shorter *J. marginatus* (grassleaf rush), which grows to 0.25 to 0.5m. Variety of open wet areas.

Juncus coriaceus - Leathery Rush

Mtn/Pdmt-FACW/CP-FACW   C values: 6 - 5 - 5

Few fruited, comparatively large rounded capsules, non-terminal inflorescences well below stem tip. Non-septate leaves. Swamp forests, wet marshes, pocosin edges, ditches.

Common *Juncus* species

Juncus dichotomus - Forked Rush

Mtn/Pdmt-FACW/CP-FACW   C values: 4 - 4 - 4

Terminal inflorescence with short branches; non-septate leaves. Variety of moist, open habitats. Occurs in drier places than most other *Juncus* species.

Juncus effusus - Common Rush

Mtn/Pdmt-FACW/CP-OBL   C values: 3 - 3 - 3

Cylindrical, non-septate leaves, often numerous within clump; multi-branched inflorescence emerging from side of flowering stem. Open marshes, shorelines, ditches, ponds.

Juncus marginatus - Grassleaf Rush

Mtn/Pdmt-FACW/CP-FACW   C values: 4 - 5 - 5

Flattened, non-septate leaves; rounded bunches of fruits scattered in terminal inflorescences. Open shores, depressions, meadows in moist sandy or peaty soil.

MONOCOT HERBS

Juncus repens - Lesser Creeping Rush

Mtn/Pdmt-OBL/CP-OBL   C values: 6 - 5 - 5

Short plant, with simple, spiky flowers, creating mats. Non-septate leaves flattened but flowering stems cylindrical. Blackwater impoundments, ditches, ponds, Carolina bays.

Juncus roemerianus - Black Needlerush

Mtn/Pdmt-n/a CP-OBL   C value: 7

Stiff, sharp-tipped needle-like blades; inflorescence not terminal. Brackish marsh species, often in extensive colonies.

Juncus scirpoides - Needlepod Rush

Mtn/Pdmt-FACW/CP-FACW   C values: 6 - 5 - 5

Terminal inflorescences of ball-shaped seedheads. Uppermost leaf longer/equal to sheath length; hollow cylindrical, septate leaves. Pocosin edges, wet savannas, marshes.

Iris virginica — Virginia Iris

Leaves emerge in fan shaped array

Beautiful, lavender or bluish-purple flowers

Often clumped; spreading by rhizomes

Emerging bright green, flattened/fanned at base

Seeds housed in elongated capsules

MONOCOT HERBS

Iridaceae - Iris Family

National Wetland Plant List
Mtn/Pdmt: **OBL**   CP: **OBL**

Coefficient of Conservatism
Mtn: **7**   Pdmt: **7**   CP: **7**

Iris virginica
Virginia Iris

Toxic Plant

Habit
Perennial herb reaching 1 m. This plant is colonial, spreading by rhizomes.

Leaves
Simple and entire with acute tips. Leaves pale blue-green, up to 1 m tall and 3 cm wide and clasping at the base.

Flowers/Fruit
Typical showy, iris flower; bluish-purple with 3 petals and 3 upwardly curved sepals. Yellow markings present on petals. Blooms April/May; fruits July to September.

Field Characteristics
Typical iris appearance with showy blue or purple flowers. Roots and seeds are somewhat toxic if eaten.

Habitat/Range
Margins of streams and ponds, freshwater marshes, swamps, wet pine flats, ditches. Most common in the Coastal Plain and rare in the eastern Piedmont.

Similar Species
Emerging iris can be confused with emerging *Typha* spp. (cattail), but emerging *Typha* leaves are bunched in a cylindrical cluster. See Common Confusions section, p. 406.

Lachnanthes caroliniana Redroot

Fan-like, overlapping leaves at base

Fuzzy white and yellow flower head

Purplish or grey-brown seedheads

Often grows in colonies in open wetlands

Roots red-orange; base of leaves "bleed" red

MONOCOT HERBS

Haemodoraceae - Bloodwort Family

Lachnanthes caroliniana
Redroot

National Wetland Plant List
Mtn/Pdmt: **OBL** CP: **OBL**

Coefficient of Conservatism
Mtn: **n/a** Pdmt: **5** CP: **5**

Habit
Perennial herb with red rhizomes and roots, and a velvety seedhead on the end of flowering stems.

Leaves
Bases of bright green leaf blades grow with one side facing the stem, about half as long as flowering stem. Leaves 30 to 38 cm long and about 1.3 cm wide, except for a few small stem leaves.

Flowers/Fruit
Compact, wooly corymb flower head with small yellow flowers, aging to dark brown, globular seed casings. Blooms June to early September; fruits September to November.

Field Characteristics
Look for characteristic red roots of this plant, which run red juice when cut. Leaves bright green and stems have fine hairs near the flower/seedhead.

Habitat/Range
Wet, acid, often sandy soil of bogs, wet pinelands, ditches, marshes, edges of pocosins.

Similar Species
Without flowers or seedheads, redroot could be confused with iris, which have brown roots.

Taxonomic Note
This is the only species in the *Lachnanthes* genus. Synonym: *Lachnanthes caroliana*, probably a spelling error.

Leersia oryzoides — Rice Cutgrass

Leaf blades, sheaths marked with fine parallel lines

Light, airy flowering panicle with very thin branchlets

Lower panicle branches partly wrapped in sheath

Hairy band along stem

Leaves have small, saw-like teeth on edges, hence the name "cutgrass"

MONOCOT HERBS

Poaceae - Grass Family

Leersia oryzoides
Rice Cutgrass

National Wetland Plant List
Mtn/Pdmt: **OBL**   CP: **OBL**

Coefficient of Conservatism
Mtn: **5**   Pdmt: **4**   CP: **4**

Drawing: Sara Fish Brown/E.O. Beal

Habit
Medium to tall (1 to 1.5 m) perennial grass with weak, slender culms.

Leaves
Yellowish-green leaves, up to 1 cm wide and 20 cm long with saw-like teeth on the margins. Leaf blades and sheaths marked with fine parallel lines.

Flowers/Fruit
Terminal inflorescence loosely branching and spreading, 10 to 20 cm long. Elliptical spikelets on wavy or undulating branches of the panicle. Flowers and fruits July to October.

Field Characteristics
Sheaths and leaf blades finely marked with parallel lines and contain small saw-like teeth on edges.

Habitat/Range
Widespread in wet areas, freshwater marshes, wet pastures and ditches. Tolerates slightly brackish waters.

Microstegium vimineum
Japanese Stilt Grass

Weak, often reclining smooth stems

Flowers in fall

Grows profusely in moist, shaded areas

Roots at nodes

Stems to 1 m long, especially later in growing season

MONOCOT HERBS

Poaceae - Grass Family

National Wetland Plant List
Mtn/Pdmt: **FAC**   CP: **FAC**

Coefficient of Conservatism
Mtn: **0**   Pdmt: **0**   CP: **0**

Microstegium vimineum
Japanese Stilt Grass

Non-native

Drawing: Karen Kendig

Habit
Low annual with freely branching, slender trailing culms, rooting at the nodes. Stems 0.5 to 1 m long.

Leaves
Leaf blades lance-shaped, 3-8 cm long and 0.5 to 1 cm wide.

Flowers/Fruit
Spikelets 0.5 cm long and in racemes of 2-6. Flowering and fruiting in fall (September to November).

Field Characteristics
Multiple spikelets distinctive. Leaves have shiny midribs. Many bottomlands are now dominated by this aggressive grass.

Habitat/Range
This introduced species inhabits floodplains, shaded banks, roadsides, marsh edges, and waste areas.

Similar Species
May be confused with *Arthraxon hispidus* (small carpgrass), but stems are smoother, leaves are finer and narrower.

Taxonomic Note
Synonym: *Eulalia vimineum*

Murdannia keisak — Wart-Removing-Herb

Leaves come off stems at right angles

Small, 3-petaled lavender flowers

This non-native can cover large areas

Fruits oblong, appearing in fall

Late in growing season, elongated stems recline in a tangle

MONOCOT HERBS

Commelinaceae - Dayflower Family

National Wetland Plant List
Mtn/Pdmt: **OBL**   CP: **OBL**

Coefficient of Conservatism
Mtn: **0**   Pdmt: **0**   CP: **0**

Murdannia keisak
Wart-Removing-Herb

Non-native

Drawing: Karen Kendig

Habit
Trailing herb forming dense mats. Stems often root at nodes.

Leaves
Alternate, linear or lance-shaped leaves, about 6 cm long and 1 cm wide. Bases of leaves have closed, tubular, hairy sheaths.

Flowers/Fruit
Lavender flowers with 3 petals and 3 sepals. Flowering and fruiting August to October.

Field Characteristics
Grass-like appearance with trailing stems.

Habitat/Range
Margins of streams, ponds, and marshes in the Coastal Plain and Piedmont of NC.

Similar Species
Sometimes looks similar to *Commelina* spp. (dayflower) but clasping hairy sheath formed at leaf base helps distinguish it.

Taxonomic Note
Synonym: *Aneilema keisak*

Peltandra virginica — Green Arrow Arum

Triangular, arrowhead shaped leaves with parallel veins

Yellow flowers on a fleshy spike

Robust herb usually found in standing water

Emerging petioles

Flowers May/June; fruits soon after flowering; fruits are pods with large green or black berries

MONOCOT HERBS

Araceae - Arum Family

National Wetland Plant List
Mtn/Pdmt: **OBL**   CP: **OBL**

Coefficient of Conservatism
Mtn: **7**   Pdmt: **7**   CP: **7**

Peltandra virginica
Green Arrow Arum

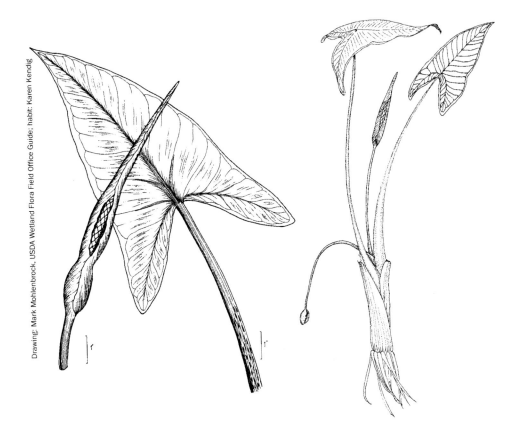

Habit
Emergent perennial herb with thick root stock. Plants about 0.5 m tall, usually found in shallow water.

Leaves
Triangular, 3-nerved (with 3 major veins), on long stems. Leaves 10-40 cm long.

Flowers/Fruit
Inconspicuous, yellow flowers on a fleshy spike (spadix), surrounded by a fleshy leaf-like structure (spathe). Fruits are pods with large green or black berries. Flowers in May/June; fruits soon after flowering.

Field Characteristics
Triangular leaves with parallel side veins, radiating from the sides of the major veins.

Habitat/Range
Bogs, freshwater marshes and perimeters of lakes and ponds. Found throughout NC, except in the northwest Mountains.

Similar Species
Can be confused with *Sagittaria latifolia* (broadleaf arrowhead), which has major leaf veins all radiating from one point at the base of the leaf. See Common Confusions section, p. 407.

Phragmites australis — Common Reed

Terminal inflorescence, long silky hairs on spikelets

No long hairs at leaf base, as in *Arundinaria gigantea*

Fence-like growth in large colonies

Dry plumes fluff out in fall and winter

This non-native can dominate marshes and lake edges, to the exclusion of many other species

MONOCOT HERBS

Poaceae - Grass Family

Phragmites australis
Common Reed

National Wetland Plant List
Mtn/Pdmt: **FACW** CP: **FACW**

Coefficient of Conservatism
Mtn: 0   Pdmt: 0   CP: 0

Non-native

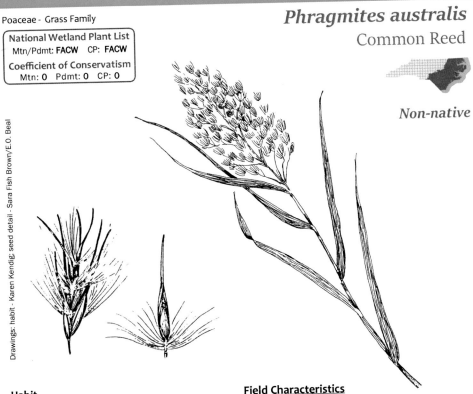

Drawings: habit - Karen Kendig; seed detail - Sara Fish Brown/E.O. Beal

Habit
Tall, perennial grass with upright culms, 2-4 m tall, forming dense and exclusive stands from creeping rhizomes.

Leaves
Broad, flat, linear blades, 1-5 cm wide and 15-40 cm long. Leaves arranged in 2 planes on stems. Stems round, thick and hollow.

Flowers/Fruit
Terminal inflorescence a densely branched panicle, 15-45 cm long with tan to purplish (when young), long, silky hairs. Spikelets contain several flowers with long, silky hairs below each flower. Flowers and fruits August to October.

Field Characteristics
Large reed with silky, dense inflorescence that tends to droop toward one side. *Phragmites* is from the Greek word, 'phragma' referring to its fence-like growth. This noxious weed now dominates many coastal marshes along the Atlantic coast.

Habitat/Range
Fresh, brackish, and salt marshes, banks of lakes and streams. *Phragmites* is worldwide in distribution and tolerates varying salinities from freshwater to saltwater.

Similar Species
Phragmites australis (common reed) can be confused with *Spartina cynosuroides* (big cordgrass) since they both occur in much the same habitats, but the seedheads differ. See Common Confusions section, p. 408. Also compare *P. australis* to the less common *Zizania aquatica* (annual wild-rice), which has a similar but more spreading seedhead. Without inflorescences, *P. australis* can be confused with *Arundinaria gigantea* (giant cane), which has long hairs extending from leaf bases, unlike *P. australis*. *A. gigantea* very rarely has inflorescences, whereas *P. australis* commonly does. Immature inflorescences are similar to *Saccharum giganteum* (sugarcane plumegrass), which has very fluffy mature plumes. *P. australis* inflorescences have long arching branches, not as compact as *Saccharum giganteum*.

Taxonomic Note
Synonym: *Phragmites communis*

Pontederia cordata — Pickerelweed

Spikes of blue, tubular flowers

Rounded, heart-shaped leaves

Generally one leaf per stem, rising from clump at base

Sometimes in dense colonies

Clump forming perennial, to 1 m

MONOCOT HERBS

Pontederiaceae - Pickerelweed Family

Pontederia cordata
Pickerelweed

National Wetland Plant List
Mtn/Pdmt: **OBL**   CP: **OBL**

Coefficient of Conservatism
Mtn: **7**   Pdmt: **5**   CP: **6**

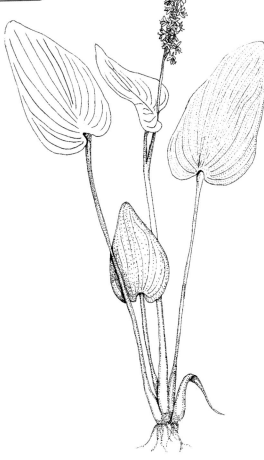

Drawing: Karen Kendig

Habit
Clumping perennial herb, to 1 m tall.

Leaves
Simple, entire, elongate, cordate leaves, 7-18 cm long.

Flowers/Fruit
Spikes of showy, blue, tubular flowers. Blooms May through October; fruits soon after flowering.

Field Characteristics
Elongated, heart-shaped leaves; base of flowering stalk is surrounded by a leaf base.

Habitat/Range
Muddy shores and shallow waters of ponds, lakes and ditches throughout NC, except in the northern Mountains.

Similar Species
Can be confused with *Sagittaria latifolia* (broadleaf arrowhead), which often grows in the same places as *Pontederia cordata* (pickerelweed). Heart-shaped leaves help distinguish *P. cordata*. See Common Confusions section, p. 407. The leaves have no midrib as does *Peltandra virginica* (green arrow arum).

Common *Rhynchospora* species

Cyperaceae - Sedge Family

Rhynchospora spp.
Beaksedge or Beakrush

Typical *Rhynchospora* appearance

Common genus of sunny, Coastal Plain marshes

Seedheads often pointed like beaks or lances

Habit
Perennial grass-like sedge with or without rhizomes.

Leaves
Linear leaves usually shorter than flowering stems. Stems smooth and may be triangular or round in cross-section.

Flowers/Fruit
Inflorescence usually in a cyme. Spikelets vary depending on the species, but often lance-shaped or beak-like.

Field Characteristics
Beak-like spikelets on inflorescences. Identification to species requires examination of mature seeds and fruits. Godfrey and Wooten (1978) drawings are very helpful.

Habitat/Range
Occur in wetland areas statewide, but most species are common only in the Coastal Plain.

MONOCOT HERBS

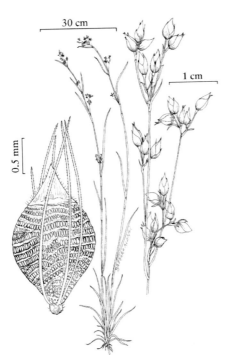

Rhynchospora caduca

Rhynchospora corniculata

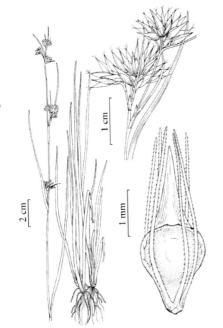

Rhynchospora glomerata

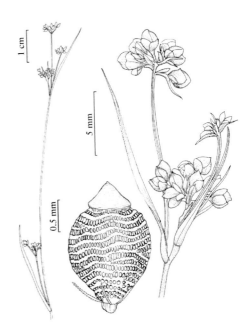

Rhynchospora globularis

Common *Rhynchospora* species

Rhynchospora caduca - Angle-stem Beaksedge

Mtn/Pdmt-OBL/CP-OBL   C values: 8-6-6

A tall, drooping sedge, to 1 m, with clusters of rounded seeds spaced along flowering stem. Open marshes, swamps, pine wetlands, impoundment shores.

Rhynchospora corniculata - Short-bristle Horned Beaksedge

Mtn/Pdmt-OBL/CP-OBL   C values: Absent-5-5

A tall, straight sedge, with large, spreading, multi-branched inflorescences on thick stem. Seeds elongated and pointed. Open freshwater marshes, swamps, freshwater shorelines, beaver ponds.

MONOCOT HERBS

Rhynchospora globularis - Globe Beaksedge

Mtn/Pdmt-FACW/CP-FACW   C values: Absent-6-5

A small, low-growing sedge, with clusters of rounded seeds at flowering stem tips. *Rhynchospora recognita* was recently split from *Rhynchospora globularis*, and looks similar but grows to 1m tall, has wider leaves and longer seeds. Pine wetlands, depressions, ditches.

Rhynchospora glomerata - Clustered Beaksedge

Mtn/Pdmt-OBL/CP-OBL   C values: 6-5-5

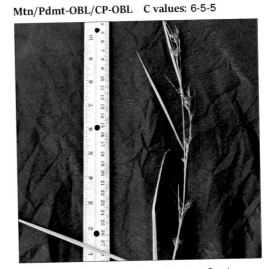

A tall, drooping sedge, with clusters of elongated, pointed seeds spaced along flowering stem. Seeds have long bristles. Pine wetlands, pocosin edges, seeps, beaver wetlands, roadside ditches.

Saccharum giganteum — Sugarcane Plumegrass

Pink-tinged seedheads appear in fall; stem hairy below

Mature seedheads open to fluffy plumes when dry

Stem leaves with long, white hairs at bases; bearded nodes

Untwisted bristles extending from mature seeds

White stripe down center of leaves; some purplish leaves

MONOCOT HERBS

Poaceae - Grass Family

National Wetland Plant List
Mtn/Pdmt: **FACW**   CP: **FACW**
Coefficient of Conservatism
Mtn: **4**   Pdmt: **4**   CP: **4**

Saccharum giganteum
Sugarcane Plumegrass

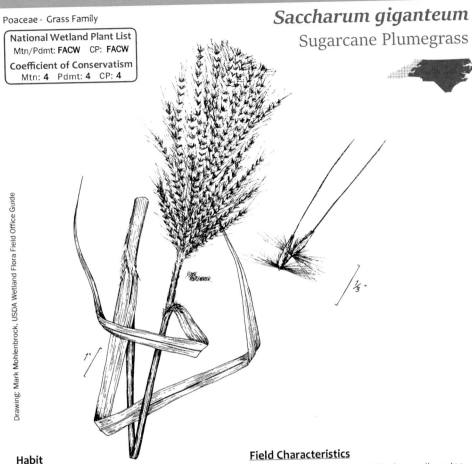

Habit
Very tall, robust, perennial grass, with large spike inflorescence on stem reaching high above leaves; flowering stems 3 - 4 m tall.

Leaves
Long blade leaves, 15-40 cm long and 2.5 cm wide, rough to the touch when rubbed from tip to base, and along edges. Stem leaf bases and young stem nodes have long, white hairs ("bearded").

Flowers/Fruit
Spikelets on the large inflorescences separate when first appearing, then aging to a fluffy silvery tan or purplish plume (when dry). Plume may be 35 cm long and 7 cm wide. Long bristles extending from mature seeds are untwisted. Flowering and fruiting late August through October.

Field Characteristics
This grass is distinguished by its huge, silvery-tan plume spikes and untwisted bristles extending from seeds. Stems below the inflorescence are hairy. Difficult to identify before seedheads appear in fall.

Habitat/Range
Fresh and brackish marshes, ditches, edges of lakes and swamps. Chiefly lower Piedmont and Coastal Plain; occasionally in Mountains.

Similar Species
Immature inflorescences are similar to *Phragmites australis* (common reed), which does not have fluffy plumes. *P. australis* has long arching inflorescence branches and is not as compact as *Saccharum giganteum* (sugarcane plumegrass). Similar to *Saccharum coarctatum* (compressed plumegrass), which is found in the same types of habitats (Coastal Plain mainly), but has a smooth stem below the inflorescence.

Taxonomic Note
Synonym: *Erianthus giganteus*

Sacciolepis striata — American Cupscale

Elongated, cylindrical inflorescence

Anthers visible during flowering

Will form dense stands

Fine stripes on inflated sacs and on leaves

Grows along sunny edges of lakes, streams, ponds, ditches, and freshwater marshes

MONOCOT HERBS

Poaceae - Grass Family

Sacciolepis striata
American Cupscale

National Wetland Plant List
Mtn/Pdmt: **OBL** CP: **OBL**

Coefficient of Conservatism
Mtn: **5** Pdmt: **4** CP: **4**

Drawing: Karen Kendig

Habit
Medium height, perennial, aquatic or semi-aquatic grass with trailing stems that root at nodes. Often forms dense stands, 0.5 to 1 m tall.

Leaves
Leaf blades flat and marked with fine parallel lines. Leaf bases cordate and clasping. Upper leaves often pointing downward.

Flowers/Fruit
Inflorescence elongate-cylindrical with spikelets on short, uneven length stalks. Spikelets distinctive with an inflated sac-like base. Flowering and fruiting July through October.

Field Characteristics
Fine stripes on leaves and on inflated 'sacs' of inflorescences are distinctive.

Habitat/Range
Sunny freshwater marshes, edges of streams, lakes and ponds, swamps, and ditches mainly in the Coastal Plain.

Sagittaria lancifolia — Bull-tongue Arrowhead

Usually found in clumps; reaches 1 m

Sunny, wet places in the Coastal Plain; brackish tolerant

Somewhat frilly flower petals

Flowers in whorls of 3 along flowering stem

Lance-shaped leaves

MONOCOT HERBS

Alismataceae - Water Plantain Family

National Wetland Plant List
Mtn/Pdmt: **OBL** CP: **OBL**

Coefficient of Conservatism
Mtn: **n/a** Pdmt: **n/a** CP: **5**

Sagittaria lancifolia
Bull-tongue Arrowhead

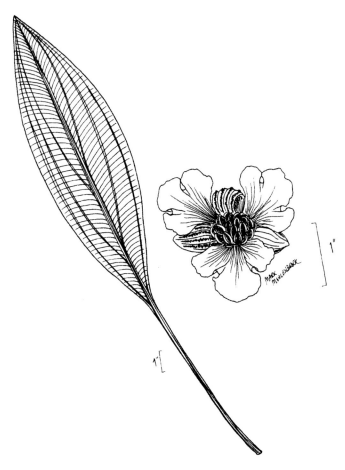

Drawing: Mark Mohlenbrock, USDA Wetland Flora Field Office Guide

Habit
Emergent or submerged perennial herb usually found in clumps, reaching about 1 m in height.

Leaves
Lance-shaped leaves with pointed tips and very long petioles.

Flowers/Fruit
White, somewhat frilly, 3-petaled flowers with yellow anthers, in whorls of 3 at nodes on flowering stalks. Blooms June to September; fruits soon after flowering.

Field Characteristics
Leaf veins radiate from petiole attachment point.

Habitat/Range
Tidal freshwater marshes or low-salinity brackish marshes, interdunal marshes and ponds, across the coast.

Taxonomic Note
Synonym: *Sagittaria falcata*
Our North Carolina variety is *Sagittaria lancifolia* var. *media* (Weakley 2020)

Sagittaria latifolia — Broadleaf Arrowhead

Flower petals less frilly than *Sagittaria lancifolia*

Ball shaped fruits, whorled in sets of three on stem

Prefers sunny wet soil, statewide

Few long veins with short connecting veins

Often found in colonies

MONOCOT HERBS

Alismataceae - Water Plantain Family

Sagittaria latifolia
Broadleaf Arrowhead

National Wetland Plant List
Mtn/Pdmt: **OBL**   CP: **OBL**

Coefficient of Conservatism
Mtn: **5**   Pdmt: **4**   CP: **5**

Drawing: Karen Kendig

Habit
Emergent or submersed, perennial herb usually found in clumps or colonies, reaching about 1 m in height.

Leaves
Triangular, with pointed tips, up to 25 cm long. Leaves have long, main veins with many short connecting veins. Petioles 5-sided in cross section.

Flowers/Fruit
White, 3-petaled flowers with yellow anthers, in whorls of 3 at nodes on flowering stalks. Blooms June to September; fruits soon after flowering.

Field Characteristics
Leaf veins radiate from petiole attachment point. Petiole 5-sided in cross section.

Habitat/Range
Wet soil, marshes, stream sides, ditches, and pond margins throughout North Carolina.

Similar Species
Sagittaria latifolia (broadleaf arrowhead) can be confused with *Pontederia cordata* (pickerelweed), which often grows in the same places and has more rounded, heart-shaped leaves. *S. latifolia* leaf veins radiate from a single point, rather than with side veins as in *Peltandra virginica* (green arrow arum), which has similar arrowhead-shaped leaves. See Common Confusions section, p. 407.

Schoenoplectus tabernaemontani — Softstem Bulrush

Loose, drooping inflorescence

Soft, bluish-green stems with no leaves

Spikelets compact with overlapping scales

Inflorescences located just below stem tip

Usually grows in standing water

MONOCOT HERBS

Cyperaceae - Sedge Family

Schoenoplectus tabernaemontani
Softstem Bulrush

National Wetland Plant List
Mtn/Pdmt: **OBL** CP: **OBL**

Coefficient of Conservatism
Mtn: 5 Pdmt: 5 CP: 5

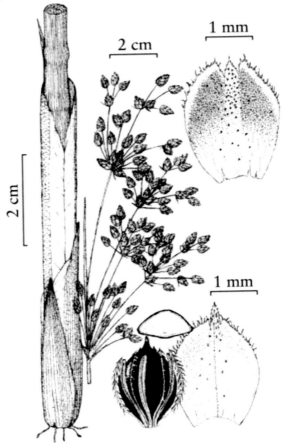

Drawing: Courtesy of the Flora of North America Association, Susan A. Reznicek and Elizabeth Zimmerman

Habit
Very tall, erect perennial, to 3 m, predominantly flowering stems; forms dense colonies.

Leaves
Leaves nonexistent or minimal sheaths to 10 cm long. Plant mostly long, grayish-green stems with single branching seedhead at the top.

Flowers/Fruit
Loose-branching umbel, widely spreading and somewhat drooping, just below stem tip. Chestnut-brown mature individual spikelets in irregular clusters, on stems of varying lengths. Each spikelet compact, with overlapping scales. Flowers and fruits June to September.

Field Characteristics
This rush has soft cylindrical spongy stems and widely spreading, drooping clusters of seeds, one per stem.

Habitat/Range
Usually found in deeper standing water, in sunny fresh and brackish marshes, muddy shores.

Similar Species
Similar to *Scirpus cyperinus* (woolgrass bulrush), but individual budlike spikelets are not as large and hairy as in *S. cyperinus*. Similar to other *Scirpus* species, but *Schoenoplectus tabernaemontani* (softstem bulrush) has no noticeable leaves.

Taxonomic Note
Synonym: *Scirpus validus*

Scirpus cyperinus — Woolgrass Bulrush

Bears a multitude of spikelets on many stalks

Strongly clumping perennial, often in standing water

Long, flexible stalks hold numerous brown spikelets

Loose branching umbel

Often grows in colonies; sunny freshwater marshes and wet meadows

MONOCOT HERBS

Cyperaceae - Sedge Family

Scirpus cyperinus
Woolgrass Bulrush

National Wetland Plant List
Mtn/Pdmt: **FACW** CP: **OBL**

Coefficient of Conservatism
Mtn: **4** Pdmt: **3** CP: **3**

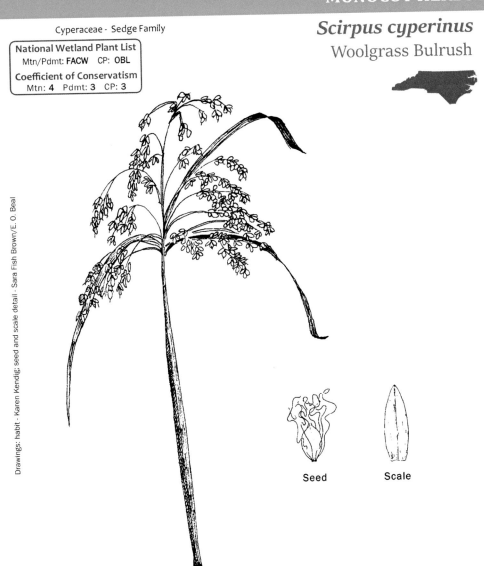

Seed   Scale

Habit
Medium to tall, erect, perennial, grass-like plant, usually 2 m tall, growing in dense clumps.

Leaves
Simple, linear leaves starting as sheaths, then extending away from stem, and drooping at the tips, up to 60 cm long. Edges rough to the touch.

Flowers/Fruit
Loose branching umbel with long flexible stalks holding brownish flowering spikelets, to 15 cm long. Rust-colored spikelets numerous; bristles much longer than seeds. Flowers and fruits July to October.

Field Characteristics
Stem may be round or weakly triangular, especially near base. Underside of leaf sheaths purple spotted.

Habitat/Range
One of the most common *Scirpus* species. Sunny freshwater marshes, wet meadows.

Similar Species
Similar to *Schoenoplectus tabernaemontani* (softstem bulrush), but *Scirpus cyperinus* (woolgrass bulrush) has noticeable leaves. Similar to *Scirpus expansus* (woodland bulrush) in the Mountains, which has reddish lower sheaths and short seed bristles.

Scirpus expansus — Woodland Bulrush

Stems rough near flowering umbel

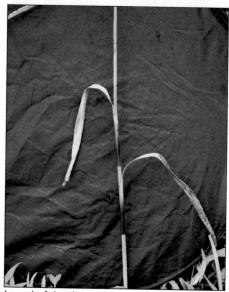
Lower leaf sheaths are red or purplish

Stalks at center of inflorescence are stiff

Leaf sheath detail

Frequent in mountain marshes, bogs, wet meadows, and sunny low spots

MONOCOT HERBS

Cyperaceae - Sedge Family

National Wetland Plant List
Mtn/Pdmt: **OBL** CP: **OBL**

Coefficient of Conservatism
Mtn: 5 Pdmt: 5 CP: n/a

Scirpus expansus
Woodland Bulrush

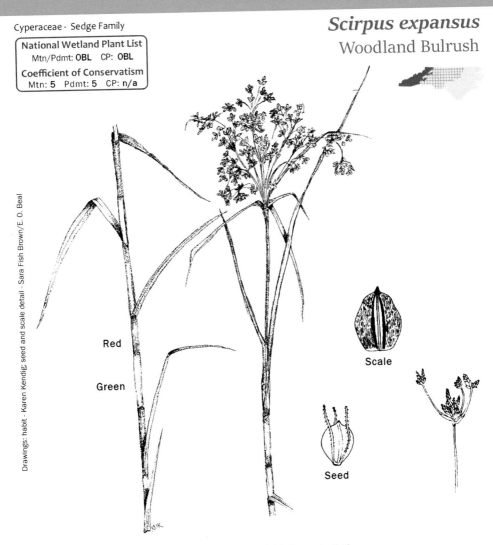

Habit
Medium to tall, erect, perennial, grass-like plant, 0.4 to 1.2 m tall.

Leaves
Leaves in sheaths and blades, separating from flowering stem in intervals nearly to the flowering head; to 60 cm long. Stems rough near flowering umbel.

Flowers/Fruit
Loose branching umbel at tip of flowering stem, with several long leaf-like bracts at its base. Stiff main branches of umbels hold secondarily branching stalks with spikelets, singly or in small groups. Scales on mature spikelets blackish. Flowers and fruits July to September.

Field Characteristics
Leaf sheaths along stems have short cross partitions, looking like joints, and the lower ones are red or purplish, a key feature.

Habitat/Range
Frequent in mountain (uncommon in upper Piedmont) marshes, bogs, wet meadows, and sunny low spots along rivers and streams.

Similar Species
Similar to *Scirpus cyperinus* (woolgrass bulrush) which lacks the reddish lower sheaths and spikelets are held on more drooping stalks. *Schoenoplectus tabernaemontani* (softstem bulrush) has a similar inflorescence but no noticeable leaves and is much taller.

Scirpus georgianus   Georgia Bulrush

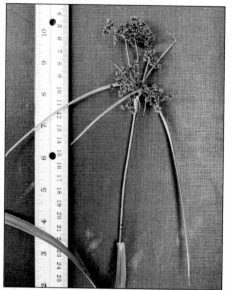
Inflorescence with few stalks, with bunches of spikelets

Perennial sedge, 0.8 m to 1.5 m tall

Spikelets a rich brown color

Spikelet detail

Rounded spikelets appear in tight bunches

MONOCOT HERBS

Cyperaceae - Sedge Family

National Wetland Plant List
Mtn/Pdmt: **OBL**   CP: **OBL**

Coefficient of Conservatism
Mtn: **5**   Pdmt: **5**   CP: **4**

Scirpus georgianus
Georgia Bulrush

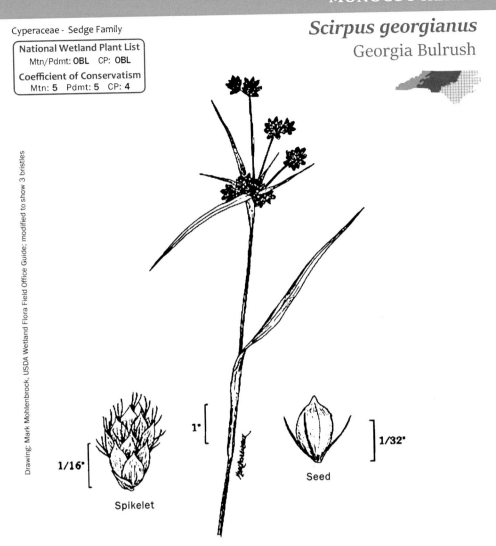

Habit
Perennial sedge, 0.8 m to 1.5 m tall, generally 1 m tall with few stem leaves.

Leaves
Mostly basal blades, to 0.5 m long, and not as numerous as in *Scirpus cyperinus* (woolgrass bulrush).

Flowers/Fruit
Seeds appear in small bunches at ends of stalks emerging from tip of flowering stem; usually an upper set and lower set, separated by erect stalks. Flowers and fruits June to September.

Taxonomic Note
The species has historically been lumped with *Scirpus atrovirens* (green bulrush), which probably does not occur in North Carolina.

Field Characteristics
Identification to species often requires examination of mature seeds. This species has 0 to 3 bristles on the seeds that are much shorter than the seed body. Secondary clusters of spikelets on very short stalks, compared to other common *Scirpus* species.

Habitat/Range
Common in the Piedmont; uncommonly in the Mountains and Sandhills. Grows in freshwater marshes, impoundment shores, ditches, wet depressions in pine areas.

Sisyrinchium angustifolium — Narrowleaf Blue-eyed Grass

Blue, 6-petaled flower; about 2 cm wide

Flowering stems are strongly flat-widened

Low, spreading perennial

Small, rounded capsules on long, drooping stalks

Capsules turn brown when mature

MONOCOT HERBS

Iridaceae - Iris Family

Sisyrinchium angustifolium
Narrowleaf Blue-eyed Grass

National Wetland Plant List
Mtn/Pdmt: **FACW** CP: **FACW**

Coefficient of Conservatism
Mtn: **4** Pdmt: **4** CP: **4**

Drawing: Britton, N.L., and A. Brown/USDA-NRCS PLANTS Database

Habit
Low, spreading perennial with strongly flat-widened, branching flowering stems and blue flowers; about 25 to 30 cm high.

Leaves
Linear blades, smooth margins, 2.5 to 5 mm wide; flowering stems branched past the middle and flowering with one or a few flowers at the tips.

Flowers/Fruit
Blue, 6-petaled flowers, 2 cm wide, on long stalks at stem ends, which are flattened. Blooms March to June; fruits June to August. Brown-rounded capsules 4 to 6 mm long.

Field Characteristics
Difficult to identify without flowers, but blue flowers on long stalks and brown capsules are distinctive.

Habitat/Range
Found in a wide array of habitats including moist woodlands, meadows, floodplain forest margins, savannas. Frequent to common throughout the state.

Similar Species
Similar to *Sisyrinchium atlanticum* (eastern blue-eyed grass), which lives in drier places and has narrower and lighter green leaves, and black fruit capsules. Flowers stalks are much shorter in most other *Sisyrinchium* species.

Sparganium americanum — American Bur-reed

Male flowering heads

Seeds produced by female flowers

Female flowers

Rounded leaf tips

Bright green leaves; bases somewhat spongy

MONOCOT HERBS

Sparganiaceae - Bur-reed Family

National Wetland Plant List
Mtn/Pdmt: **OBL**   CP: **OBL**
Coefficient of Conservatism
Mtn: **7**   Pdmt: **6**   CP: **6**

Sparganium americanum
American Bur-reed

Drawing: Mark Mohlenbrock, USDA Wetland Flora Field Office Guide

Nutlet

Habit
Medium height perennial, up to 1 m tall, with thick, spongy leaves and spiky, ball-like fruits.

Leaves
Mostly basal, simple, entire, thick and spongy. Leaves often grow taller than flowering stems. Leaf tips somewhat rounded.

Flowers/Fruit
Ball-shaped flowering heads, born sequentially along a zig-zag flowering stem; female flowers on lower balls and male flowers on upper balls. Fruits are balls comprised of nutlets. Flowers and fruits from May to September.

Field Characteristics
Leaf tips rounded; round ball-like flowers/fruits distinctive, with male and female flowers on separate balls.

Habitat/Range
Often in standing water in marshes, sunny edges of slow streams, beaver ponds, swamps, and lakes; statewide, but chiefly in the Coastal Plain and Mountains.

Spartina alterniflora — Smooth Cordgrass

Usually growing in water in dense stands

Flowering stalks

Leaves concave in cross section

Abundant plant in brackish and salt marshes

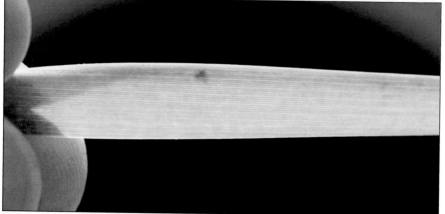
Blades smooth, or nearly so, with smooth edges

MONOCOT HERBS

Spartina alterniflora
Smooth Cordgrass

Poaceae - Grass Family

National Wetland Plant List
Mtn/Pdmt: **n/a** CP: **OBL**

Coefficient of Conservatism
Mtn: **n/a** Pdmt: **n/a** CP: **7**

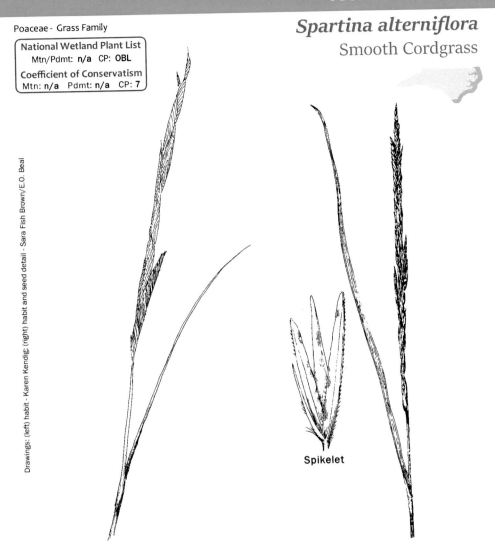

Drawings: (left) habit - Karen Kendig; (right) habit and seed detail - Sara Fish Brown/E.O. Beal

Spikelet

Habit
Medium to tall (0.5 to 2.5 m), upright, perennial grass. Contains soft, spongy culms which may be a centimeter thick at base. Plants may be shorter and stunted on higher ground.

Leaves
Long grass-like blades flat and tapered to a sharp point. Leaves about 1 cm wide and up to 40 cm long and typically smooth or nearly so.

Flowers/Fruit
Terminal inflorescence is compact so it appears cylindrical. Inflorescence about 10-30 cm long with 5-30 alternately arranged spikelets. Flowering stem tends to be one-sided. Flowers and fruits August to October.

Field Characteristics
Smooth cordgrass is the most abundant and ecologically significant large plant in brackish or salt marshes as it supplies detritus to the estuaries. Salt marshes are comprised almost solely of this species.

Habitat/Range
Salt or brackish marshes along the outer coast, frequently growing in water and forming dense stands to the exclusion of nearly all other species.

Taxonomic Note
Synonym: *Sporobolus alterniflorus*

Spartina cynosuroides — Big Cordgrass

Very large inflorescence

Pictured: Milo Pyne
Plant 2 to 3 meters tall

Rounded ridge on blade undersides; fine teeth on edges

Often on banks or slight rises along channels

Found in brackish and fresh tidal marshes; can be locally abundant

MONOCOT HERBS

Poaceae - Grass Family

Spartina cynosuroides
Big Cordgrass

National Wetland Plant List
Mtn/Pdmt: **n/a**   CP: **OBL**

Coefficient of Conservatism
Mtn: **n/a**   Pdmt: **n/a**   CP: **7**

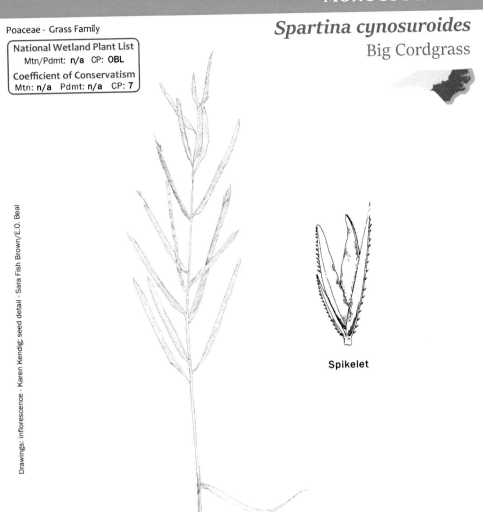

Spikelet

Habit
Tall, stout, upright perennial grass growing in stands, reaching 2– 4 m; spreading by elongated rhizomes.

Leaves
Elongated, linear tapering leaf blades, up to 70 cm long and 1 to 2.5 cm wide with scabrous margins. Stems thick, round and hollow.

Flowers/Fruit
An open, terminal panicle with many spikes ascending and spreading. Spikelets usually 10-12 cm long. Flowers and fruits June to September.

Field Characteristics
Stems thick, round, and hollow; long, wide leaf blades with rough edges and rounded ridge on underside. Very large inflorescence, but fairly sparse spikelets.

Habitat/Range
Brackish and fresh tidal marshes along the outer Coastal Plain or intermixed in marshes dominated by *Juncus roemerianus* (black needlerush).

Similar Species
Spartina cynosuroides (big cordgrass) resembles the non-native *Phragmites australis* (common reed) and occupies much of the same habitat. *P. australis* has a denser inflorescence and stiff, straight leaf blades. See Common Confusions section, p. 408.

Taxonomic Note
Synonym: *Sporobolus cynosuroides*

Spartina patens — Saltmeadow Cordgrass

Open, terminal panicle with 3 to 6 spikes

Blades cylindrical and thin; stems wiry and hollow

Grows in colonies, often leaning or looking flattened

Sparsely branching

Populations of this meadow-like grass often do not flower every year

MONOCOT HERBS

Poaceae - Grass Family

National Wetland Plant List
Mtn/Pdmt: **n/a** CP: **FACW**

Coefficient of Conservatism
Mtn: **n/a** Pdmt: **n/a** CP: **7**

Spartina patens
Saltmeadow Cordgrass

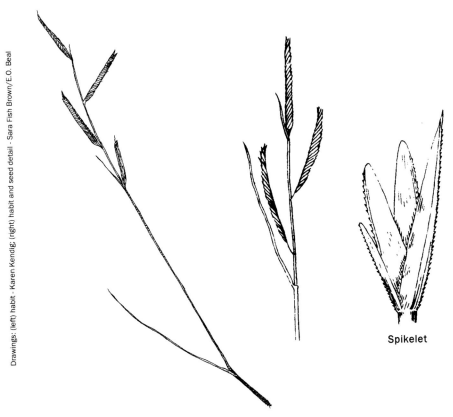

Spikelet

Habit
Fairly low to medium height, graceful, meadow-like grass, up to 1 m tall. Spreading by elongated rhizomes.

Leaves
Narrow, linear leaf blades rolled inward and less than 3 mm wide and 35 cm long. Stems wiry and hollow.

Flowers/Fruit
Open, terminal panicle with 3 to 6 alternately arranged spikes which contain densely packed spikelets, 7-12 mm long. Flowers and fruits June to September, but populations often do not flower every year.

Field Characteristics
This grass may spread by runners in straight lines or may form tufted growths.

Habitat/Range
Brackish marshes, low sand dunes and sand flats along the outer Coastal Plain. Saltmeadow cordgrass can grow in vast expanses above the high tide line.

Similar Species
Spartina patens (saltmeadow cordgrass) has wiry stems whereas other species of *Spartina* have wider stems.

Taxonomic Note
Synonym: *Sporobolus pumilus*

Typha angustifolia — Narrowleaf Cattail

Very narrow, dark green leaves

Long, thin cattail

Note space between male and female flowers

Stem cross-section; leaves strongly convex

Found in outer Coastal Plain; fresh to brackish areas

MONOCOT HERBS

Typhaceae - Cattail Family

National Wetland Plant List
Mtn/Pdmt: n/a   CP: OBL

Coefficient of Conservatism
Mtn: n/a   Pdmt: n/a   CP: 0

Typha angustifolia
Narrowleaf Cattail

Drawing: Karen Kendig

Habit
Perennial herb 1-3 m tall.

Leaves
Dark green, narrow leaves which are strongly convex in cross section, up to 1 cm wide.

Flowers/Fruit
Dark brown cattail spike with a space between female and male flower spikes. Leaves, which come off lower part of stem, stand taller than flowering stem. Flowers May to July; fruits June to November.

Field Characteristics
Look for a space between the male and female cattail spikes and narrow, dark leaves, up to 1 cm wide.

Habitat/Range
Brackish to slightly brackish tidal marshes, ditches, and pond/lake margins in the outer Coastal Plain.

Similar Species
Similar to *Typha latifolia* (broadleaf cattail), but *T. angustifolia* (narrowleaf cattail) has darker green, narrower leaves and much narrower cattail spikes. Leaves in cross section are more convex than in *T. latifolia*.

Taxonomic Note
This cattail species is native in Eurasia, and also the northeast and mid-Atlantic coast of North America. Not native inland.

Typha latifolia — Broadleaf Cattail

No space between male and female flowers

Cattail prop roots

Emerging cattail can be confused with emerging iris

Less convex than *Typha angustifolia* in cross-section

Freshwater cattail; statewide, but most common in Piedmont. Fluffy seedheads remain through winter.

MONOCOT HERBS

Typhaceae - Cattail Family

National Wetland Plant List
Mtn/Pdmt: **OBL**   CP: **OBL**

Coefficient of Conservatism
Mtn: **2**   Pdmt: **2**   CP: **2**

Typha latifolia
Broadleaf Cattail

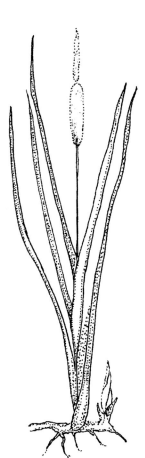

Drawing: Karen Kendig

Habit
Tall, perennial herb, 2-3 m high.

Leaves
Light green leaves arising from the sheathing base. Strap shaped leaves to 2.5 cm wide and very long, 2 to 2.5 m long.

Flowers/Fruit
Brown, cylindrical terminal spike (female flower), familiar to most people. When present, male flower spike positioned adjacent and above female flower spike on flowering stem. Flowers May to July; fruits June to November.

Field Characteristics
Leaves flat or slightly convex near base. Emerging cattail can be confused with emerging iris (*Iris* spp.), but iris leaves emerge in a fan shape. Cattail leaves emerge in a cylindrical shape.

Habitat/Range
Common in freshwater marshes, ditches and ponds. Found statewide, although most common in the Piedmont.

Similar Species
Similar to *Typha angustifolia* (narrowleaf cattail), a Coastal Plain species. *T. latifolia* (broadleaf cattail) has wider, bluish-green, less convex leaves and a larger spike than *T. angustifolia*. Male flower spike is directly above female flower spike on *T. latifolia*, whereas *T. angustifolia* has a space between male and female spikes. Emerging *Typha* spp. can be confused with emerging *Iris* spp.; see Common Confusions section, p. 406.

Xyris spp. — Yellow-eyed Grass

Leaves attached at base; emerging in fan shape

Yellow petaled flowers on cone-like spikes

Mature cone-like seedheads

Sometimes multiple flowers at once

Frequently in colonies

MONOCOT HERBS

Xyridaceae - Yellow-eyed Grass Family

Xyris spp.
Yellow-eyed Grass

National Wetland Plant List
Mtn/Pdmt: **Most OBL** CP: **Most OBL**
Coefficient of Conservatism
Varied, nearly all 6 - 9

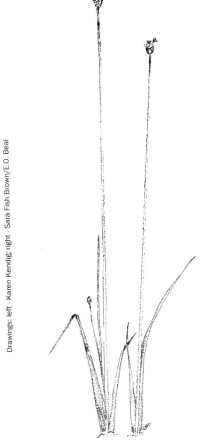

Drawings: left - Karen Kendig; right - Sara Fish Brown/E.O. Beal

Habit
Medium or small height, perennial herb, with narrow linear leaves all emerging from a fibrous or bulbous base.

Leaves
Leaves linear, flat blades, often twirling, attached at base.

Flowers/Fruit
Three yellow, rounded petals born on solitary, brown, cone-like spikes with overlapping scales. Fruit is capsule with tiny seeds. Generally flowering and fruiting June to September, but varies widely by species.

Field Characteristics
Recognizable by flattened, thick, upright leaf blades, often twirling; leaves of individual species vary considerably in size, from 5 cm to 1 m. Identification to species requires examination of mature seeds and fruit capsules.

Habitat/Range
Sunny spots in standing water at pond edges, wet sandy ditches, marshes, pine wetlands. More common in the Coastal Plain, but a few species are found in other ecoregions.

Zephyranthes atamasca Atamasco Lily

Pink-tinged buds

Big, showy, white (rarely pink) flowers

Shiny black seeds in swollen capsules, split at maturity

Leaves arise from underground bulb

Long, linear, basal leaves

MONOCOT HERBS

Amaryllidaceae - Amaryllis Family

Zephyranthes atamasca
Atamasco Lily

National Wetland Plant List
Mtn/Pdmt: **FACW**   CP: **FACW**

Coefficient of Conservatism
Mtn: **7**   Pdmt: **7**   CP: **6**

Highly Toxic Plant

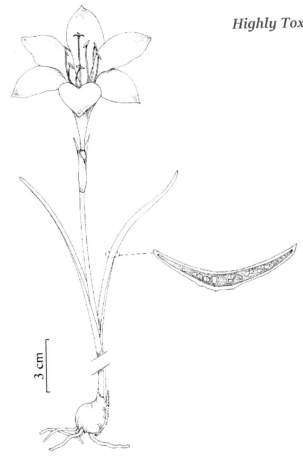

Drawing: Courtesy of the Flora of North America Association, Yevonn Wilson-Ramsey

Habit
Low, perennial herb with narrow leaves, flowering in spring.

Leaves
Cluster of shiny, blade-like, basal leaves, cupped in cross-section, about 30 cm long and 0.5 cm wide.

Flowers/Fruit
Hollow, leafless flowering stalk about 30 cm tall, with single, large white, 6-petaled lily flower, turning light pink with age. Fruit is swollen capsule, housing shiny, black seeds, and splitting at maturity. Flowers March/April; fruits May/June.

Field Characteristics
Flower is not aromatic. May be fatal if eaten.

Habitat/Range
Shady bottomland forests, depressions, wet meadows, damp roadsides in the eastern Piedmont and northwestern Coastal Plain.

Taxonomic Note
Synonyms: *Atamosco atamasca*, *Zephyranthes atamasco*

DICOT HERBS

Scientific Name	Common Name	National Wetland Plant List Status Mtns & Piedmont/Coastal Plain	Page
Alternanthera philoxeroides	Alligatorweed	OBL/OBL	282
Asclepias incarnata	Swamp Milkweed	OBL/OBL	284
Bidens frondosa	Devil's Beggarticks	FACW/FACaW	286
Boehmeria cylindrica	False Nettle	FACW/FACW	288
Centella erecta	Erect Centella	(n/a)/FACW	290
Cicuta maculata	Spotted Water Hemlock	OBL/OBL	292
Cuphea carthagenensis	Colombian Waxweed	(n/a)/FAC	294
Drosera spp.	Sundew species	OBL/OBL	296
Eclipta prostrata	False Daisy	FAC/FACW	298
Galium tinctorium	Bedstraw	OBL/FACW	300
Gratiola virginiana	Round-fruit Hedge-hyssop	OBL/OBL	302
Hydrocotyle umbellata	Marsh Pennywort	OBL/OBL	304
Hypericum mutilum	Dwarf St. John's Wort	FACW/FACW	306
Impatiens capensis	Jewelweed	FACW/FACW	308
Limonium carolinianum	Carolina Sea-lavender	(n/a)/OBL	310
Lindernia dubia	Yellow-Seed False Pimpernel	OBL/OBL	312
Lobelia cardinalis	Cardinalflower	FACW/FACW	314
Ludwigia alternifolia	Seedbox	FACW/OBL	316
Ludwigia palustris	Marsh Primrose-Willow	OBL/OBL	318
Lycopus virginicus	Virginia Water Horehound	OBL/OBL	320
Mimulus spp.	Monkey-flower species	OBL/OBL	322
Penthorum sedoides	Ditch Stonecrop	OBL/OBL	324
Persicaria sagittata	Arrowleaf Tearthumb	OBL/OBL	326
Pilea pumila	Canadian Clearweed	FACW/FACW	328
Pluchea foetida	Stinking Camphorweed	OBL/OBL	330
Pluchea odorata	Sweetscent	FACW/FACW	332
Polygonum spp.	Smartweed species	Varies/Varies	334
Ptilimnium capillaceum	Mock Bishopweed	OBL/OBL	336
Ranunculus abortivus	Kidney-Leaf Buttercup	FACW/FACW	338
Ranunculus recurvatus	Blisterwort	FAC/FACW	340
Rhexia mariana	Maryland Meadowbeauty	OBL/FACW	342
Salicornia spp.	Glasswort species	OBL/OBL	344
Saururus cernuus	Lizard's Tail	OBL/OBL	346
Solidago patula	Roundleaf Goldenrod	OBL/OBL	348
Solidago sempervirens	Seaside Goldenrod	(n/a)/FACW	350
Symphyotrichum pilosum	Frost Aster	FAC/FAC	352
Verbena urticifolia	White Vervain	FAC/FAC	354

Alternanthera philoxeroides — Alligatorweed

Flowers and fruits March to October

Small, globular, white flowers attached at leaf base

Stems to 1 m long, looking jointed where leaves attach

Leaves wider past the middle

Forms extensive floating mats

Amanda Mueller

DICOT HERBS

Amaranthaceae - Amaranth Family

National Wetland Plant List
Mtn/Pdmt: **OBL** CP: **OBL**

Coefficient of Conservatism
Mtn: **n/a** Pdmt: **0** CP: **0**

Alternanthera philoxeroides
Alligatorweed

Non-native

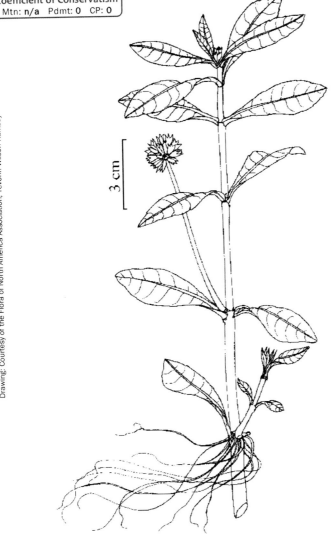

Drawing: Courtesy of the Flora of North America Association, Yevonn Wilson-Ramsey

Habit
Vine-like, weedy perennial, often growing in dense floating mats, rooted in water up to 2 m deep.

Leaves
Opposite, toothless, smooth and somewhat succulent leaves. Stems hollow.

Flowers/Fruit
Small, white, globular flowers born on long stalks emerging from base of leaves. Flowers and fruits March-October.

Field Characteristics
Stems can be up to 1 m long, look jointed where leaves attach, and are red-tinged at the joints.

Habitat/Range
In mats on still waters of fresh to brackish rivers, canals, ditches, ponds, and lake margins of the Coastal Plain and Piedmont.

Asclepias incarnata — Swamp Milkweed

Leafy, 1 to 1.25 m tall herb

Generally unbranched until the top

Erect seed pods contain seeds with tufts of long hairs

Milkweeds exude white sap

Few other marsh plants have rose-colored umbels

DICOT HERBS

Asclepiadaceae - Milkweed Family

National Wetland Plant List
Mtn/Pdmt: **OBL**   CP: **OBL**

Coefficient of Conservatism
Mtn: **5**   Pdmt: **5**   CP: **5**

Asclepias incarnata
Swamp Milkweed

Toxic Plant

Drawing: Sara Fish Brown/E.O. Beal

Habit
A leafy 1 to 1.25 m tall herb, generally unbranched until the top.

Leaves
Opposite, lance-shaped, about 10 to 12 cm long.

Flowers/Fruit
Many small clusters of bright pink flowers at branch ends. Blooms July to September; fruits August to October.

Field Characteristics
All milkweeds have milky sap which may irritate skin; few other marsh plants have umbels of rose-colored flowers. A favorite of butterflies and other insects.

Habitat/Range
Sunny marshes mainly, but also mountain bogs, wet meadows, openings in swamps, primarily in the Mountains, Piedmont, and the northeastern Coastal Plain. Absent from the southeastern quarter of the state.

Taxonomic Note
Common name is a misnomer - this milkweed is a marsh species.

Bidens frondosa — Devil's Beggarticks

Leaf-like bracts extend from base of yellow flowers

Annual herb, to 1 m tall, many branching

Yellow disk and ray flower

Seeds with two barbed awns

Flowers and fruits June to October

DICOT HERBS

Asteraceae - Aster Family

National Wetland Plant List
Mtn/Pdmt: **FACW**   CP: **FACW**
Coefficient of Conservatism
Mtn: **2**   Pdmt: **2**   CP: **2**

Bidens frondosa
Devil's Beggarticks

Habit
Medium annual herb, to 1 m tall; often many branching.

Leaves
Leaf blades lance shaped, toothed, usually divided into 3 to 5 leaflets.

Flowers/Fruit
Yellow disk and ray flowers usually have only disks, 30 to 60 or more. Leaflike bracts (about 8) extend from base of flowers. Barbed, flat seeds have two barbed awns, "devil's pitchforks". Flowers and fruits June to October.

Field Characteristics
Look for yellow disk flowers with green leaf-like bracts extending from base, or the flat, barbed seeds.

Habitat/Range
Wet meadows, marshes, floodplain forests, ditches, beaver marshes, and waste places across the state, except in the Outer Banks.

Boehmeria cylindrica — False Nettle

Perennial herb to 1 m

Leaves generally dull, not glossy

Outer main veins curve inside toothed leaf margins

Unbranched inflorescences attach to stem

Often grows in colonies

DICOT HERBS

Urticaceae - Nettle Family

Boehmeria cylindrica
False Nettle

National Wetland Plant List
Mtn/Pdmt: **FACW**   CP: **FACW**

Coefficient of Conservatism
Mtn: **5**   Pdmt: **4**   CP: **4**

Drawing: Karen Kendig

Habit
Perennial herb to 1 m.

Leaves
Opposite, sometimes sub-opposite, broad, lance-shaped leaves with toothed edges. Leaves dull, not glossy. Outer veins curve inside leaf margins.

Flowers/Fruit
Spikes of small, spherical clusters borne in leaf axils. Blooms July/August, and fruits September/October.

Field Characteristics
Not irritating compared to the similar stinging nettle.

Habitat/Range
Common in low ground, swamps and wet woods throughout NC.

Similar Species
Appears similar to *Pilea pumila* (Canadian clearweed), although flowers of *Boehmeria cylindrica* (false nettle) are in long spikes compared to clearweed's short branched panicles; *P. pumila* leaves are glossy. See Common Confusions section, p. 408.

Centella erecta — Erect Centella

With competition, leaves are more erect

Leaves pubescent front and back

Without competition, *Centella* lies closer to the ground

Tiny, round, flattened fruits

Tiny, 5-petaled, white flowers

DICOT HERBS

Apiaceae - Parsley Family

National Wetland Plant List
Mtn/Pdmt: **n/a**   CP: **FACW**

Coefficient of Conservatism
Mtn: **n/a**   Pdmt: **n/a**   CP: **4**

Centella erecta
Erect Centella

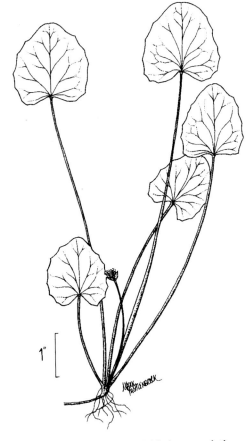

Habit
Low-growing, creeping, perennial herb, clusters of leaves arising from nodes.

Leaves
Somewhat heart-shaped, basal leaves have long petioles of variable lengths, to 30 cm. Margins generally dentate, but variable between leaves or plants. Leaves and flower stalks usually hairy.

Flowers/Fruit
Tiny, 5-petaled, white flowers in umbels, on short stalks emerging from base. Fruits tiny, flattened-round pods. Blooms June to August; fruits July to September.

Field Characteristics
Flowering stalks hairy and much shorter than leaf petioles.

Habitat/Range
Found in sunny pond, lake, and stream edges, ditches, wet grasslands, and a wide variety of other moist to wet habitats in the Coastal Plain.

Similar Species
Hydrocotyle umbellata (marsh pennywort) has petioles attached at the center of disk-shaped leaves; see Common Confusions section, p. 409.

Taxonomic Note
Synonym: *Centella asiatica* (misapplied); our plants are now understood to be a related species native to North America.

Cicuta maculata — Spotted Water Hemlock

Tall, late branching herb to 2 m tall

Sturdy stems often have purplish hue

White-flowered umbels, divided into sub-umbels

Strongly toothed leaves

Lance-shaped leaflets with deeply serrated margins

DICOT HERBS

Apiaceae - Parsley Family

Cicuta maculata
Spotted Water Hemlock

National Wetland Plant List
Mtn/Pdmt: **OBL** CP: **OBL**

Coefficient of Conservatism
Mtn: **6** Pdmt: **5** CP: **5**

Highly Toxic Plant

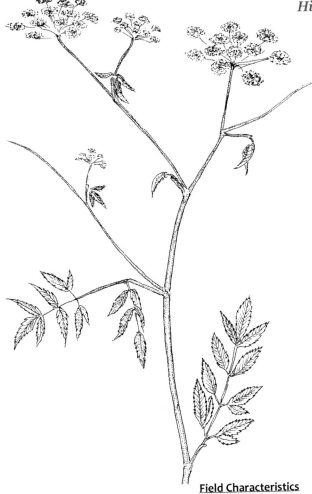

Drawing: Karen Kendig

Habit
Tall, late-branching herb, 1.5 to 2 m tall, with divided, serrated leaves and white umbel flowers.

Leaves
Alternate, divided into numerous lance-shaped leaflets with deeply serrated margins. Stems often purplish.

Flowers/Fruit
Usually numerous, rounded, white-flowered umbels, composed of sub-umbels. Blooms May to August; fruits July to September.

Field Characteristics
Extreme caution should be used when handling this plant, as it is highly poisonous if ingested. This is the most common tall white umbel-flowering wetland species east of the Mountains.

Habitat/Range
Marshes, bogs, wet meadows, ditches, swamp and bottomland openings across the state.

Cuphea carthagenensis — Colombian Waxweed

Small to medium herb, 30 to 60 cm tall

Opposite, wide, elliptical leaves with short petioles

Hairy stems

Fruit a tiny capsule with yellowish-brown seeds

Flower petals have a darker purple streak in center

DICOT HERBS

Lythraceae - Loosestrife Family

National Wetland Plant List
Mtn/Pdmt: **n/a** CP: **FAC**

Coefficient of Conservatism
Mtn: **n/a** Pdmt: **0** CP: **0**

Cuphea carthagenensis
Colombian Waxweed

Non-native

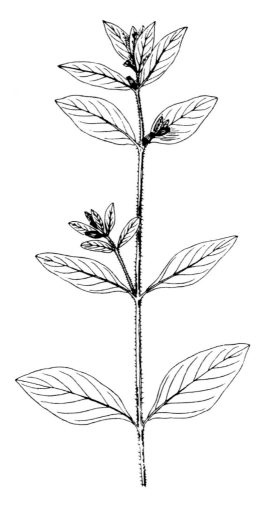

Drawing: Sara Fish Brown/E.O. Beal

Habit
Small to medium herb, 30 to 60 cm, with few upward branches and wide, elliptic leaves.

Leaves
Opposite, wide, elliptic with very short petioles. Hairy stems.

Flowers/Fruit
Small flowers with a green or reddish tube and six fragile, purple petals with darker center stripes. Fruit capsule with yellowish-brown seeds. Flowering and fruiting June to September.

Field Characteristics
Small plant with hairy stems and leaves; red center stripe on petals of small flowers.

Habitat/Range
Marshes, ditches, wet meadows, shallow water of floodplain forests, swamps, and depressions throughout the Coastal Plain.

Taxonomic Note
Has been called *Cuphea carthagensis*, a spelling error

Drosera spp. — Sundew

Drosera intermedia - long petioles, leaf blades longer than wide

Drosera brevifolia - large flower on glandular stalk

Drosera intermedia

Drosera intermedia - flower stalk not glandular

Drosera capillaris turns red in full sun. Fruit are capsules (*D. capillaris* fruit - right).

DICOT HERBS

Droseraceae - Carnivorous Family

Drosera spp.
Sundew

National Wetland Plant List
Mtn/Pdmt: **OBL**   CP: **OBL**

Coefficient of Conservatism
Mtn: **8**   Pdmt: **7**   CP: **7**

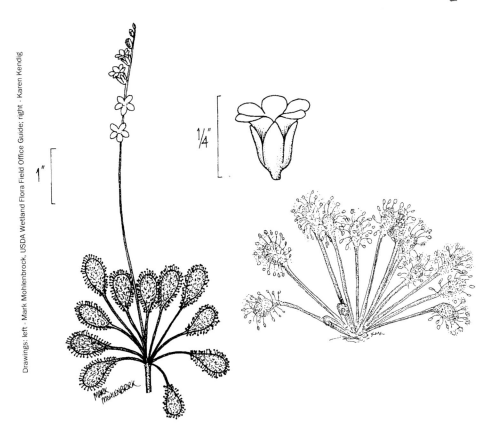

Drawings: left - Mark Mohlenbrock, USDA Wetland Flora Field Office Guide; right - Karen Kendig

Habit
Small, low-growing carnivorous plant that traps insects in drops of sticky secretions on hairy leaves.

Leaves
Alternate, entire. Upper surfaces and margins of leaves covered with tentacle-like hairs that secrete drops of a sticky substance to trap insects.

Flowers/Fruit
White or pink flowers are one-sided racemes on stalks rising above leaves; individual 5-petaled flowers opening one at a time. Fruits are capsules with tiny seeds. Bloom April through September, fruiting soon after.

Field Characteristics
Sundews are unmistakable, with their "dewdrop" covered round leaves.

Habitat/Range
Drosera capillaris (pink sundew), *D. intermedia* (spoonleaf sundew), and *D. brevifolia* (dwarf sundew) are common in the Coastal Plain, in wet sandy places, pine wetlands, and pond edges; *D. rotundifolia* (roundleaf sundew) is more rare and found in mountain bogs, fens, and seeps.

Similar Species
Five species occur in North Carolina.

Eclipta prostrata   False Daisy

Many short, thin petals on flattened flowers

Occurs in sunny wet edges

Long, linear or lance-shaped, slightly toothed leaves

Grows easily in disturbed wet areas

Leaves at right angles to stems

DICOT HERBS

Asteraceae - Aster Family

Eclipta prostrata
False Daisy

National Wetland Plant List
Mtn/Pdmt: **FAC**   CP: **FACW**

Coefficient of Conservatism
Mtn: **2**   Pdmt: **2**   CP: **2**

Drawing: Sara Fish Brown/E.O. Beal

Nutlet

Habit
Annual herb, semi-erect with stems rooting at nodes, usually less than 60 cm long.

Leaves
Opposite, long linear or lance-shaped, slightly toothed leaves. Leaves to 13 cm long.

Flowers/Fruit
Small, white and light yellow flowers, with many short, thin, white petals on ray flowers. Bracts longer than petals. Disk flowers develop into nutlets that are quadrangular on top. Flowering and fruiting June to November.

Field Characteristics
Sometimes appears weedy. Flowers are distinctive.

Habitat/Range
Sunny edges of rivers, oxbow ponds, wet ditches, low woods, swamps, and wet disturbed areas. Statewide, but more common in the Coastal Plain and eastern Piedmont; rare in the upper Piedmont and Mountains.

Taxonomic Note
Synonym: *Eclipta alba*

Galium tinctorium — Bedstraw

Leaves whorled along weak stems

Fruits are tiny, smooth, spherical pods

Stems and leaves slightly rough to the touch

Sometimes appears as a sprawling tangle

Flowers have three or four petals

DICOT HERBS

Rubiaceae - Bedstraw Family

National Wetland Plant List
Mtn/Pdmt: **OBL**   CP: **FACW**

Coefficient of Conservatism
Mtn: 6   Pdmt: 5   CP: 5

Galium tinctorium
Bedstraw

Habit
Medium height weak herb, with long, rough stems and whorled, small, linear leaves.

Leaves
Simple, narrow leaves in whorls of 5 or 6 (rarely 4), widely spaced on square stems. Leaves elongated and blunt or rounded at tips, about 1 to 3 mm wide. Leaves and stems slightly rough to the touch.

Flowers/Fruit
Three-petaled, small, white flowers in clusters usually of 3, arising from bases of leaves; fruits are tiny, smooth, spherical pods, ripening to black. Blooms April to June; fruits June to August.

Field Characteristics
Weakly erect or in tangled, reclining strands, stems somewhat branching. Look for variably-sized whorled leaves in sets of 5 and 6, and tiny, 3-petaled white flowers.

Habitat/Range
Swamps, wet meadows and ditches, bogs, marshes statewide.

Similar Species
This bedstraw species differs from the others by having 3 or 2-petaled flowers.

Gratiola virginiana — Round-fruit Hedge-hyssop

White, tubular flowers on short stalks at leaf bases

Round, capsule fruit

Thick, fleshy stems; succulent leaves

Commonly in wet mud or very shallow water

Generally reclining, unbranched herb

DICOT HERBS

Scrophulariaceae - Figwort Family

National Wetland Plant List
Mtn/Pdmt: **OBL**   CP: **OBL**
Coefficient of Conservatism
Mtn: 5   Pdmt: 5   CP: 5

Gratiola virginiana
Round-fruit Hedge-hyssop

Habit
Short, unbranched, reclining herb, 15 to 20 cm long, with thick stems.

Leaves
Opposite, elliptic to lance-shaped leaves, usually serrated, about 4 cm long and 1 cm wide and thick.

Flowers/Fruit
Small, white, tubular flowers on short stalks at the bases of leaves. Fruit round, at leaf bases. Blooms mainly March to May, fruiting soon after.

Field Characteristics
Thick, fleshy stem; succulent leaves. Generally reclining.

Habitat/Range
Common in the Coastal Plain and Piedmont in wet mud or very shallow water. Found along pond shores, ditches, pools, and sunny openings in swamps.

Similar Species
Lindernia dubia (yellow-seed false pimpernel) leaves are wide at the base, unlike *Gratiola virginiana* (round-fruit hedge-hyssop).

Hydrocotyle umbellata — Marsh Pennywort

Round, disk-shaped leaves with scalloped edges

Flowers are umbels on separate stems from leaves

Also called manyflower marsh-pennywort

Leaves have no cuts in round disk

Petioles attach to center of disk leaf

DICOT HERBS

Araliaceae - Ginseng Family

Hydrocotyle umbellata
Marsh Pennywort

National Wetland Plant List
Mtn/Pdmt: **OBL**   CP: **OBL**

Coefficient of Conservatism
Mtn: **n/a**   Pdmt: **4**   CP: **4**

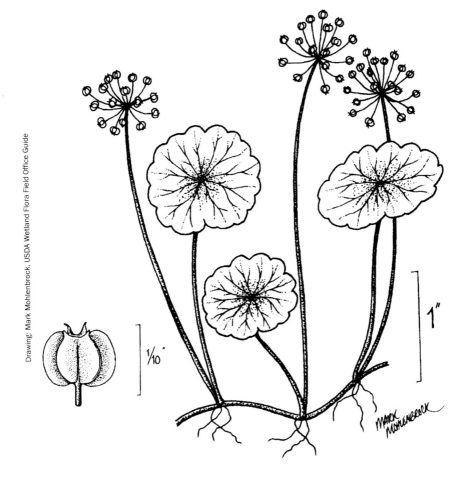

Habit
Low-growing, creeping, semi-aquatic, perennial herb, 15 - 25 cm high, rooted at nodes; sometimes in floating mats.

Leaves
Alternate, simple, round leaves with scalloped margins and long petioles 4 - 15 cm long.

Flowers/Fruit
Small, white flowers in simple umbels, with stalks as long as leaf petioles. Blooms April to September, fruits soon after.

Field Characteristics
Distinctive disk-shaped leaves with petiole attached in the center. Flowering stalks are smooth and long.

Habitat/Range
Fresh and brackish marshes, shaded shores of ponds and lakes, ditches, swamps. More common in the Coastal Plain than in the Piedmont.

Similar Species
Centella erecta (erect centella) has petioles that are not attached at the center of their heart-shaped leaves. See Common Confusions section, p. 409.

Hypericum mutilum — Dwarf St. John's Wort

Thin-stemmed, leafy, minimally branching herb

Tiny, five-petaled flowers with long stamens

Grows to 30 cm tall; sometimes in colonies

Leaves attach directly to stems

Flowering stems attach at base of leaves; leaves develop reddish spots with age

DICOT HERBS

Hypericaceae - St. John's Wort Family

National Wetland Plant List
Mtn/Pdmt: **FACW**   CP: **FACW**
Coefficient of Conservatism
Mtn: **4**   Pdmt: **3**   CP: **3**

Hypericum mutilum
Dwarf St. John's Wort

Drawing: Karen Kendig

Habit
Thin-stemmed, leafy, minimally branching herb, to 30 cm tall.

Leaves
Opposite, ovate to elliptic leaves, rounded at base, up to 4 cm long and 1.5 cm wide. Leaves paler green beneath, directly attaching to stem. Leaves develop reddish spots with age.

Flowers/Fruit
Many tiny, 5-petaled, yellow flowers with long stamens, mostly at tips of branches. Yellow seeds contained in capsules. Flowering June to October; fruiting soon after.

Field Characteristics
Small, but recognizable by its rounded, sessile leaves.

Habitat/Range
Common throughout the state in open sunny wet or damp areas like bogs, marshes, ditches, pond margins, and low areas.

Similar Species
Compare to *Hypericum punctatum* (spotted St. John's wort), which is a larger plant with larger flowers. Petals and leaves of *H. punctatum* have evident oil glands.

Impatiens capensis — Jewelweed

Often in colonies; blooms all summer until frost

Smooth, succulent stems

Medium to tall bright green herb, reaching 1.5 m

Tubular, orange, spotted flowers

Mature fruit is elongated capsule which explodes with seeds when touched

DICOT HERBS

Balsaminaceae - Touch-me-not Family

National Wetland Plant List
Mtn/Pdmt: **FACW**   CP: **FACW**
Coefficient of Conservatism
Mtn: **4**   Pdmt: **4**   CP: **4**

Impatiens capensis
Jewelweed

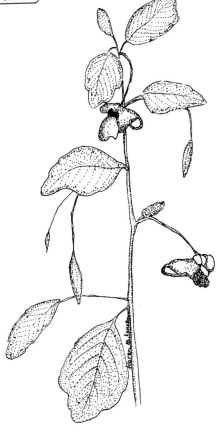

Drawing: Karen Kendig

Habit
Medium to tall, bright green, succulent herb, reaching 1.5 m.

Leaves
Smooth, succulent stems contain alternately arranged ovate to elliptic leaves produced on long petioles. Leaves 3-10 cm long and 3-8 cm wide with crenate edges.

Flowers/Fruit
Orange, tubular "bugle-like" flowers with brownish spots. Ends of flowers have curved spurs. Mature football-shaped seed capsules explode and scatter seeds when ripe, hence the plant's other common name, "touch-me-not." Blooms from May to frost; fruits soon after flowering.

Field Characteristics
Watery fluids of jewelweed are thought to dilute the sap of *Toxicodendron radicans* (eastern poison ivy), making the rash less severe.

Habitat/Range
Stream and lake edges and moist woods throughout NC.

Similar Species
A similar species with yellow flowers, *Impatiens pallida* (pale jewelweed), occurs in wet woods and seepage areas in the mountains.

Limonium carolinianum — Carolina Sea-lavender

Airy, branching inflorescence above fleshy basal leaves

Tiny purple flowers

Branching inflorescence

Often mixed with other salt marsh plants

Fleshy basal leaves

DICOT HERBS

Plumbaginaceae - Plumbago Family

National Wetland Plant List
Mtn/Pdmt: **n/a** CP: **OBL**

Coefficient of Conservatism
Mtn: **n/a** Pdmt: **n/a** CP: **8**

Limonium carolinianum
Carolina Sea-lavender

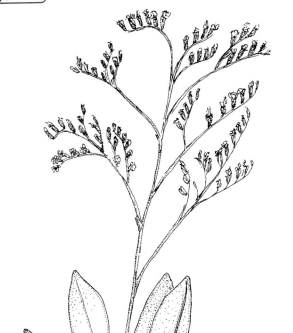

Drawing: Karen Kendig

Habit
Woody roots, leathery, basal leaves; tall, branching, open inflorescence with purple flowers.

Leaves
Fleshy or leathery, simple, alternate with thick petioles clasping the base of the plant. Leaves lance or spoon-shaped with prominent mid-rib.

Flowers/Fruit
Small, 5-petaled, lavender or purple flowers in a large, airy, branching panicle, fan-shaped in profile, to 0.6 to 1 m tall. Blooms August to October, fruiting soon after.

Field Characteristics
Look for fleshy leaves in a basal rosette; branching floral stalk with tiny purple flowers.

Habitat/Range
Strongly saline conditions: salt marshes, interdunal swales, salt flats.

Lindernia dubia — Yellow-Seed False Pimpernel

Flowers emerge singly from bases of leaves

Weakly ascending, low annual herb, 10 to 25 cm tall

Leaves attach directly to stem

Tiny lavender flowers

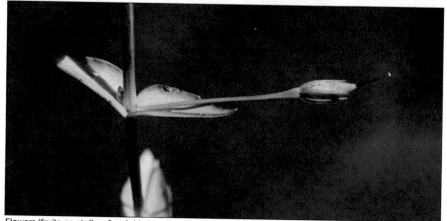
Flowers/fruits on stalks of variable lengths, depending on variety

DICOT HERBS

Scrophulariaceae - Figwort Family

National Wetland Plant List
Mtn/Pdmt: **OBL**   CP: **OBL**

Coefficient of Conservatism
Mtn: **5**   Pdmt: **4**   CP: **4**

Lindernia dubia
Yellow-Seed False Pimpernel

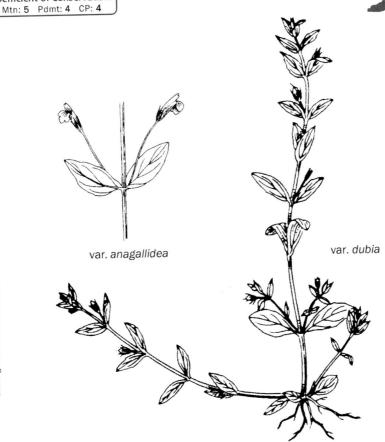

var. *anagallidea*

var. *dubia*

Drawing: Sara Fish Brown/E.O. Beal

Habit
Weakly ascending, low, annual, leafy herb, 10 to 25 cm tall.

Leaves
Opposite leaves in scattered pairs, about 3 cm long and 1 cm wide, obovate, tapered to the stem, sometimes toothed.

Flowers/Fruit
Small, pale purple or blue flowers, approximately 1 cm long, growing singly from leaf bases on short stalks of variable lengths. Flowers June through September, fruiting soon after.

Field Characteristics
Leaves taper to a narrow base; stems somewhat reclining.

Habitat/Range
A common species in freshwater wet places, such as stream floodplains, bottomlands, ditches, muddy lake and pond shorelines, and wet meadows.

Similar Species
Gratiola virginiana (round-fruit hedge-hyssop) has a similar habit, but fruits are round and its leaves taper to narrow bases.

Taxonomic Note
Lindernia dubia var. *dubia* has flowering stalks shorter than leaves; *L. dubia* var. *anagallidea* has flowering stalks longer than leaves.

Lobelia cardinalis — Cardinalflower

Showy, bright red flowers on tall flowering stalks

Basal rosettes persist through winter

Usually growing in standing water

Fruits are capsules

Leaves may have smooth or serrated edges; when serrated, small and large teeth alternate along the edges

DICOT HERBS

Campanulaceae - Bluebell Family

National Wetland Plant List
Mtn/Pdmt: **FACW**   CP: **FACW**

Coefficient of Conservatism
Mtn: **6**   Pdmt: **5**   CP: **5**

Lobelia cardinalis
Cardinalflower

Toxic Plant

Drawing: Karen Kendig

Habit
Medium to tall perennial herb, 0.5 m to rarely 2 m tall.

Leaves
Alternate, elliptic leaves which can be entire or serrated, 4-20 cm long and 2-3 cm wide. When serrated, small and large teeth alternate along margins.

Flowers/Fruit
Spikes of two-lipped, intensely red, tubular flowers. Spikes are usually about 20 cm long, but can reach 50 cm in length. Fruits are capsules. Blooms July to October; fruits soon after flowering.

Field Characteristics
Showy, bright red flowers blooming in mid-late summer. Basal rosettes persist through winter. Plant toxic if eaten in large quantities.

Habitat/Range
Freshwater marshes and swamps, river banks, ditches and stream sides throughout NC.

Ludwigia alternifolia — Seedbox

Flower petals fall quickly

Sepals remain long after petals fall

Erect, tall herb, 1 to 1.25 m

Distinctive box-shaped seed capsules

Reddish stem and petioles; long narrow leaves

DICOT HERBS

Onagraceae - Evening Primrose Family

Ludwigia alternifolia
Seedbox

National Wetland Plant List
Mtn/Pdmt: **FACW**   CP: **OBL**

Coefficient of Conservatism
Mtn: **4**   Pdmt: **4**   CP: **4**

Drawings: left and mature fruit detail - Karen Kendig; immature fruit detail - Sara Fish Brown/E.O. Beal

Habit
Erect, tall herb when mature, with branches turned upward, 1 to 1.25 m tall.

Leaves
Alternate, lance-shaped to narrowly elliptic with very short petioles and smooth margins. Stems reddish, angled, and slightly winged.

Flowers/Fruit
Bright yellow, 4-petaled flowers, about 1.25 cm across. Round yellow petals often drop quickly, but 4 green sepals remain. Capsule distinct, 4-angled and box-shaped, remaining into the winter. Blooms May to October, fruiting soon after.

Field Characteristics
Look for 4-petaled yellow flowers, and brown box-shaped capsules.

Habitat/Range
Quite common throughout the state in open, sunny, wet or damp areas like bogs, marshes, ditches, pond margins, low areas, and openings in swamps.

Similar Species
Other erect-growing North Carolina *Ludwigia* species tend to have more than four petals or none at all.

Ludwigia palustris — Marsh Primrose-Willow

Leaves with pointed ends

Plant may be submerged

Creeping, leafy plant with reddish stems to 60 cm long

Tiny flowers arise from bases of leaves

Tiny capsule fruits have dark and light greenish stripes

DICOT HERBS

Onagraceae - Evening Primrose Family

National Wetland Plant List
Mtn/Pdmt: **OBL**   CP: **OBL**

Coefficient of Conservatism
Mtn: **4**   Pdmt: **4**   CP: **4**

Ludwigia palustris
Marsh Primrose-Willow

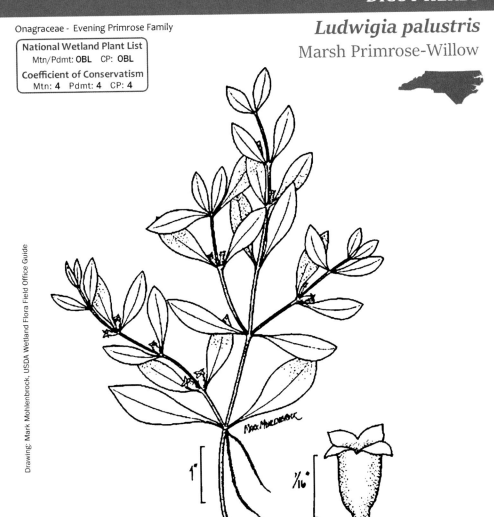

Drawing: Mark Mohlenbrock, USDA Wetland Flora Field Office Guide

Habit
Creeping, leafy plant with small leaves, rooted at nodes, growing about 60 cm long.

Leaves
Opposite, elliptic with smooth margins and pointed ends, about 3 cm long and 1.5 cm wide. Leaves taper along petiole to stems, which are red.

Flowers/Fruit
Tiny flowers with no petals (just 4 green sepals), directly attached to stems just above leaves. Fruit a tiny capsule (2 to 4 mm long) with green stripes. Blooms May to October, fruiting soon after flowering.

Field Characteristics
Roots at the nodes and stays recumbent even in maturity.

Habitat/Range
Very common throughout the state sprawled on exposed mud or in shallow water at pond or lake margins, marsh edges, openings in swamps, wet ditches.

Similar Species
Can be confused with immature *Ludwigia hexapetala* (common water-primrose), but look for leaves with pointed ends and reddish stems. Similar to *Rotala ramosior* (lowland rotala) but with much wider leaves.

Lycopus virginicus — Virginia Water Horehound

Toothed leaves with winged petioles, white center vein

Sometimes in colonies

May be purplish when young or in sun

Tiny flowers at base of leaves

Rounded fruits along four-sided stem

DICOT HERBS

Lamiaceae - Mint Family

National Wetland Plant List
Mtn/Pdmt: **OBL**   CP: **OBL**

Coefficient of Conservatism
Mtn: 5   Pdmt: 5   CP: 5

Lycopus virginicus
Virginia Water Horehound

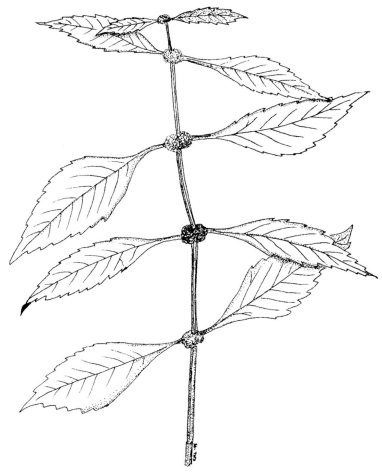

Habit
Small to medium perennial herb, less than 1 m in height.

Leaves
Opposite, toothed, dark green, lance-shaped leaves, sometimes with purple coloration.

Flowers/Fruit
Small, white, tubular flowers form distinctive spherical clusters or whorls at the base of the leaves. Blooms from July to frost; fruits soon after flowering.

Field Characteristics
Stems 4-sided; clusters of axillary flowers are distinctive.

Habitat/Range
Wet meadows, swamps, streambanks, perimeters of ponds statewide.

Mimulus spp. Monkey-flower

Mimulus alatus (flowers on short stalks)

Mimulus ringens (flowers on long stalks)

Mimulus alatus (flower on short stem, leaves with petiole)

Lavender tubular flowers

Mimulus ringens (leaves attach directly to stem, flowers on long stalks)

DICOT HERBS

Phrymacea - Lopseed Family

National Wetland Plant List
Mtn/Pdmt: **OBL**   CP: **OBL**

Coefficient of Conservatism
Mtn: **5/6**   Pdmt: **5/6**   CP: **5/6**

Mimulus spp.
Monkey-flower

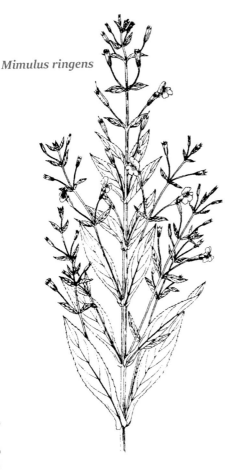

Mimulus ringens

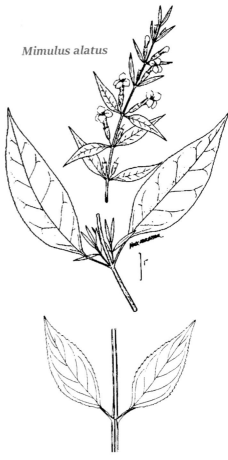

Mimulus alatus

Habit
Tall, branching, leafy herb, to 2 m.

Leaves
Mimulus alatus has leaves with petioles and stems distinctly winged. *M. ringens* has sessile leaves and no wings on stems.

Flowers/Fruit
Monkey-flower has lavender tubular flowers. *Mimulus alatus* has sessile flowers and *M. ringens* has flowers on long stalks. *M. alatus* blooms July to November, fruiting in fall. *M. ringens* blooms June to September, fruiting soon after flowering.

Field Characteristics
These two species seldom grow together in the same wetland or ditch. Noting leaf arrangement and length of flower stalks helps distinguish them from each other.

Habitat/Range
Mimulus ringens (Allegheny monkey-flower; C value 6) occurs chiefly in the Mountains and Piedmont, along marsh edges, pond margins, ditches, wet meadows, and bogs. *M. alatus* (sharpwing monkey-flower; C value 5) occurs predominantly in the Coastal Plain and lower Piedmont, in ditches, marsh edges, and openings in bottomlands.

Penthorum sedoides — Ditch Stonecrop

Ditches, marshes, low wet areas, stream, pond margins

Distinctive arching inflorescence, two or more racemes

Stamens extend above flower

Fruits are lobed capsules

Finely toothed leaves

DICOT HERBS

Penthoraceae - Ditch-stonecrop Family

Penthorum sedoides
Ditch Stonecrop

National Wetland Plant List
Mtn/Pdmt: **OBL**   CP: **OBL**

Coefficient of Conservatism
Mtn: **4**   Pdmt: **4**   CP: **4**

Drawing: Courtesy of the Flora of North America Association, Barbara Alongi

Habit
Leafy, robust, branching, perennial herb, growing to about 0.75 m high.

Leaves
Alternate, narrow lance-shaped with finely toothed edges; not succulent. Leaves 5 to 10 cm long, to 4 cm wide.

Flowers/Fruit
Yellowish or whitish-green flowers, and later lobed capsules, on the upper side of two or more arching stems, at the ends of branches. Flowers June to October, fruiting soon after.

Field Characteristics
Arching flowering stems distinctive and often lingering from one season into the next.

Habitat/Range
Abundant in ditches, marshes, swamp openings, floodplain pools, and stream and pond margins throughout the state, but less common in the Mountains and the far eastern counties.

Taxonomic Note
This is the only *Penthorum* species in the western hemisphere.

Persicaria sagittata — Arrowleaf Tearthumb

Distinct arrowhead shaped leaves almost wrap stems

Small, globular flower heads on long stalks

Tiny hooked barbs line the stems

Spines sometimes continue on leaf undersides

Often in extensive colonies

DICOT HERBS

Polygonaceae - Smartweed Family

National Wetland Plant List
Mtn/Pdmt: **OBL**   CP: **OBL**

Coefficient of Conservatism
Mtn: 5   Pdmt: 4   CP: 5

Persicaria sagittata
Arrowleaf Tearthumb

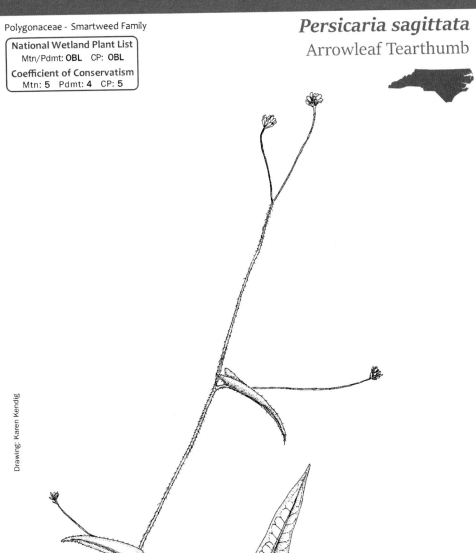

Drawing: Karen Kendig

Habit
Trailing, perennial herb up to 2 m long with weak intertwining branches.

Leaves
Simple, entire, alternate, arrow-shaped leaves with acute tips. Leaf midribs contain spines.

Flowers/Fruit
White to pink, clustered on ends of stalks emerging from base of leaves. Flowers and fruits May to November.

Field Characteristics
Square stem heavily armed with unmistakable briars that can tear flesh, hence the common name.

Habitat/Range
Freshwater marshes, wet fields and disturbed areas across NC.

Taxonomic Note
Synonym: *Polygonum sagittatum*

Pilea pumila — Canadian Clearweed

Three prominent veins, rising from bases of leaves

Flowers August/September

Branching, flowering panicles arise from leaf axils

Glossy, stout stems

Glossy leaves; annual, understory plant

DICOT HERBS

Urticaceae - Nettle Family

National Wetland Plant List
Mtn/Pdmt: **FACW** CP: **FACW**

Coefficient of Conservatism
Mtn: **4** Pdmt: **4** CP: **4**

Pilea pumila
Canadian Clearweed

Drawing: Karen Kendig

Habit
Annual herb with glossy leaves and translucent stems, up to 0.5 m.

Leaves
Opposite, ovate, toothed leaves with 3 prominent veins. Leaves shiny on upper side and 4 to 10 cm long, produced on long petioles.

Flowers/Fruit
Greenish-white flowers in a branched panicle arising from the leaf axil. Blooms August/September; fruits September to November.

Field Characteristics
On the glossy leaves, look for unbranched center vein and three curving major veins.

Habitat/Range
Often in dense colonies in wet soil or shallow freshwater margins, low, shady pastures. or moist, shady uplands across the state. Less common in southern half of the Coastal Plain.

Similar Species
Leaves are similar to *Boehmeria cylindrica* (false nettle) which grows much taller, has dull leaves, and flowers in cylindrical, unbranched spikes. See Common Confusions section, p. 408.

Pluchea foetida — Stinking Camphorweed

Unbranched, leafy perennial, to 1 m tall

Leaves attach directly to, or clasp, stem

Entire plant is strongly malodorous, like all *Plucheas*

Seeds attached to long bristles for dispersal

Tightly packed clusters of "dirty white" flowers

DICOT HERBS

Asteraceae - Aster Family

National Wetland Plant List
Mtn/Pdmt: **OBL**   CP: **OBL**

Coefficient of Conservatism
Mtn: **n/a**   Pdmt: **5**   CP: **5**

Pluchea foetida
Stinking Camphorweed

Drawing: Courtesy of the Flora of North America Association, Barbara Alongi

Habit
Unbranching, leafy perennial, 0.5 to 1 m tall, topped with dense clusters of white disk flowers.

Leaves
Opposite, clasping, fuzzy, dull, thick green leaves; stems often purplish.

Flowers/Fruit
Flat-topped clusters of white disk flowers. Seeds with long bristles for dispersal. Flowers and fruits late July to October.

Field Characteristics
Crushed leaves strongly malodorous, hence the species name. Look for clasping, fuzzy leaves.

Habitat/Range
Mainly found in the Coastal Plain and lower Piedmont, in marshes, Carolina bays, ditches, and borrow pits.

Similar Species
The less common *Pluchea camphorata* (camphor pluchea) is found in wetlands statewide. It has light pink flowers and leaves with petioles (not clasping). Also similar to *Pluchea odorata* (sweetscent), which also has short petioles (not clasping) and grows in coastal wetlands.

Pluchea odorata — Sweetscent

Tall (to 1.5 m), unbranching, leafy perennial

Entire plant strongly malodorous, like all *Plucheas*

Seeds attached to bristles for dispersal

Tightly packed, medium to dark pink flowers

Leaves do not clasp stem; short petioles

DICOT HERBS

Asteraceae - Aster Family

National Wetland Plant List
Mtn/Pdmt: **FACW**   CP: **FACW**
Coefficient of Conservatism
Mtn: **n/a**   Pdmt: **n/a**   CP: **5**

Pluchea odorata
Sweetscent

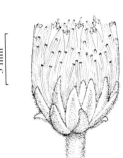

Drawing: Courtesy of the Flora of North America Association, Barbara Alongi

Habit
Unbranching, leafy perennial, to 1.5 m tall, topped with dense clusters of light pink disk flowers.

Leaves
Elliptic, thick leaves with short petioles or tapered at base (not clasping). Inflorescence stem and branches pubescent.

Flowers/Fruit
Flat-topped clusters of medium to dark pink disk flowers. Seeds attached to long bristles for dispersal. Flowers and fruits August to October.

Field Characteristics
Plants have an unpleasant fragrance; look for short petioles.

Habitat/Range
Estuarine habitats in the outer Coastal Plain: freshwater tidal, brackish, and salt marshes.

Similar Species
The less common *Pluchea camphorata* (camphor pluchea) is found in wetlands statewide. It has light pink flowers and leaves with petioles (not clasping). Similar to *Pluchea foetida* (stinking camphorweed), which has clasping leaves and grows in freshwater wetlands.

Taxonomic Note
In North Carolina, this species is restricted to coastal counties. In the far south and southwest US, this species is found well inland, despite its other common name, saltmarsh fleabane.

Polygonum spp.   Smartweed

Flowers do not open all at once

Sheaths extend upward from leaf nodes

Often grows in extensive colonies

Typical *Polygonum* spp. inflorescence

Sometimes called jointweed after the swollen joints along stems

DICOT HERBS

Polygonaceae - Smartweed Family

National Wetland Plant List
Mtn/Pdmt: **Varies**   CP: **Varies**
Coefficient of Conservatism
Varies

Polygonum spp.
Smartweed

Some non-native

Drawing: (left) Mark Mohlenbrock, USDA Wetland Flora Field Office Guide; (right) Karen Kendig

Habit
Erect or somewhat trailing herb, rooting at lower nodes.

Leaves
Alternate, and typically narrowly elliptic with leaf sheaths (ocrea) extending upward from leaf nodes.

Flowers/Fruit
Small flowers in loose terminal spikes, white to pink. Flowering and fruiting May/June to October/November.

Field Characteristics
Members of this genus contain a leaf sheath (ocrea), formed by stipules encircling the stem. The leaf sheath may be densely hairy. The bitter taste of these leaves are said to "smart", hence the common name, smartweed.

Habitat/Range
Marshes and shores of lakes and ponds throughout NC.

Taxonomic Note
Many taxonomists now place smartweeds in the genus *Persicaria*.

Ptilimnium capillaceum — Mock Bishopweed

White flowering umbels; fruits pointed oval shaped

Often reclining on other plants

Umbels divided into 10 or more umbellets

Leaves thread-like, divided several times

Tiny, white, five-petaled flowers; blooms June to August

DICOT HERBS

Apiaceae - Umbellifer Family

National Wetland Plant List
Mtn/Pdmt: **OBL**   CP: **OBL**

Coefficient of Conservatism
Mtn: **5**   Pdmt: **4**   CP: **4**

Ptilimnium capillaceum
Mock Bishopweed

Drawing: Sara Fish Brown/E.O. Beal

Habit
Fine-leaved, semi-erect herb, 30 to 60 cm tall, with tiny, white, umbel flowers.

Leaves
Leaves thread-like, divided several times.

Flowers/Fruit
Umbels of tiny white flowers, petals only about 1 mm long and wide. Fruits light brown, ribbed oval shaped. Flowers June to August; fruits July to September.

Field Characteristics
This species usually has 10 or more umbellets per flowering umbel, and 10 or more flowers per umbellet.

Habitat/Range
Open and sunny wet places, such as marshes, ditches, depressions, and swamp and bottomland openings in the Coastal Plain and lower Piedmont.

Ranunculus abortivus — Kidney-Leaf Buttercup

Low herb with tall flowering stems

Fruits soon after flowering March to June

Basal rosette of leaves

Kidney-shaped basal leaves

Tiny yellow flowers with greenish centers

DICOT HERBS

Ranunculaceae - Buttercup Family

National Wetland Plant List
Mtn/Pdmt: **FACW**   CP: **FACW**

Coefficient of Conservatism
Mtn: **3**   Pdmt: **3**   CP: **3**

Ranunculus abortivus
Kidney-Leaf Buttercup

Toxic Plant

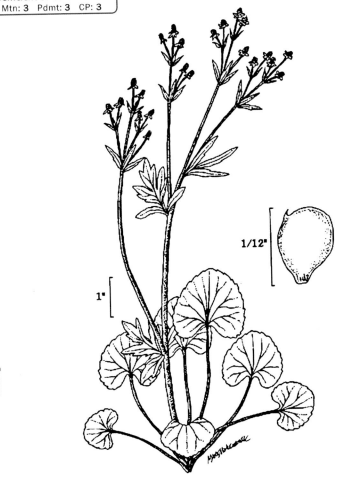

Drawing: Mark Mohlenbrock, USDA Wetland Flora Field Office Guide

Habit
Low herb with tall, often branching flowering stems.

Leaves
Distinctive basal leaves with long petioles and kidney-shaped blades about 2.5 cm wide. Flowering stems have smaller, linear leaves, widely spaced.

Flowers/Fruit
Flowering stems up to 0.5 m tall above basal leaves. Tiny yellow flowers with greenish centers later mature to green seedheads. Blooms March to June; fruits soon after flowering.

Field Characteristics
Leaf blades usually scalloped and wider than long.

Habitat/Range
Moist to wet soils in partly sunny areas of floodplains, bottomlands, swamp margins, and other low ground, even on roadsides; common over most of the state except the Coastal Plain.

Ranunculus recurvatus — Blisterwort

Long, flowering stems

Found in damp or wet woods

Three-lobed leaves

Tiny yellow and green flowers

Seeds have distinctive recurved hooks

DICOT HERBS

Ranunculaceae - Buttercup Family

National Wetland Plant List
Mtn/Pdmt: **FAC**   CP: **FACW**

Coefficient of Conservatism
Mtn: **6**   Pdmt: **5**   CP: **5**

Ranunculus recurvatus
Blisterwort

Toxic Plant

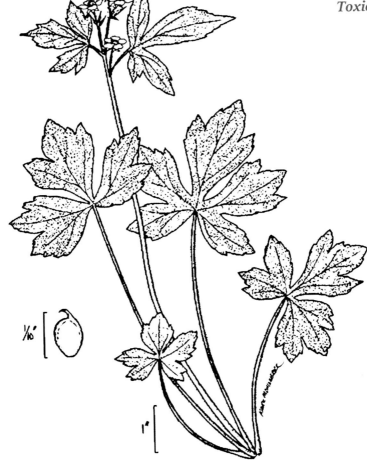

Drawing: Mark Mohlenbrock, USDA Wetland Flora Field Office Guide

Habit
Low growing herb, with tall flowering stalks, to 0.5 m tall.

Leaves
Basal leaves with long, hairy petioles. Leaves divided or lobed into three segments with serrated edges.

Flowers/Fruit
Flowering stalk quite hairy, with a few large leaves like the basal leaves. Tiny yellow flowers develop to seedheads with hooked seeds. Flowers April to June; fruits soon after flowering.

Field Characteristics
Leaves have hairy petioles. Distinctive seeds with curved hooks.

Habitat/Range
Found in the Mountains, Piedmont, and inner Coastal Plain primarily. Common in moist soil, such as bottomlands, swamps, or damp slopes.

Rhexia mariana — Maryland Meadowbeauty

Pink flowers are quick to fade

Upright, mostly unbranched plant

Unique jointed stamens on *Rhexia* spp. flowers

Pointed bud and globulose fruit

Hairy stems and leaves distinguish this *Rhexia* species. Linear, toothed leaves

DICOT HERBS

Melastomataceae - Melastome Family

National Wetland Plant List
Mtn/Pdmt: **OBL**   CP: **FACW**

Coefficient of Conservatism
Mtn: **5**   Pdmt: **4**   CP: **4**

Rhexia mariana
Maryland Meadowbeauty

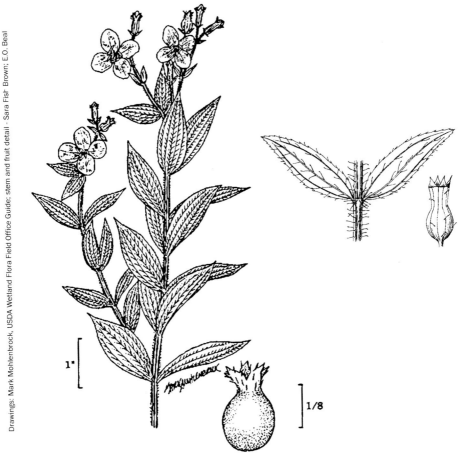

Habit
Medium height, upright, perennial herb; stems generally about 30 to 40 cm tall.

Leaves
Opposite, small, variable leaves, hairy with three parallel veins, 2.5 to 5 cm long. Leaves attach directly to stem or with a very short petiole.

Flowers/Fruit
Showy, pale or medium pink flowers, 3 cm wide, with 8 conspicuous yellow jointed stamens. Petals pale or medium pink, 2.5 cm long. Fruits urn-shaped capsules with globulose bases. Blooms May to October; fruits soon after flowering.

Field Characteristics
Flowers appear fresh in morning, but often fading by day's end, especially in heat. Four-sided, hairy stems not symmetrical, having two sides rounded and wider than the other two sides.

Habitat/Range
Ephemeral pond edges, pine wetlands, wet meadows, pocosin edges, and upper edges of wet ditches.

Similar Species
Most other *Rhexia* species in North Carolina are found just in the Coastal Plain, but *Rhexia mariana* (Maryland meadowbeauty) is common throughout.

343

Salicornia spp. — Glasswort

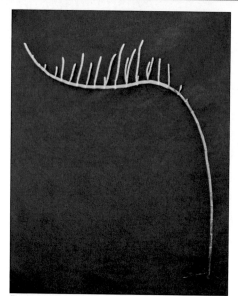
Fleshy, perennial herb, growing from a rhizome

Succulent plant found in salt and brackish marshes

Stems are green, succulent, jointed tubes

Forms colonial mats

Salicornia sp. flowering

DICOT HERBS

Chenopodiaceae - Goosefoot Family

Salicornia spp.
Glasswort

National Wetland Plant List
Mtn/Pdmt: **OBL** CP: **OBL**

Coefficient of Conservatism
Mtn: **n/a** Pdmt: **n/a** CP: **8**

Drawing: Sara Fish Brown/E.O. Beal

Habit
Upright, low-growing, fleshy, perennial herb reaching 0.3 m high, often smaller. Glasswort forms colonial mats, growing from a horizontal rhizome.

Leaves
Inconspicuous leaves reduced to small scales. Stems consist of green, succulent, jointed tubes.

Flowers/Fruit
Tiny flowers along ends of jointed tubes. Flowering and fruiting July to October.

Field Characteristics
Fleshy, succulent plant found in salt flats. Edible plant; tastes salty.

Habitat/Range
Brackish and salt flats and marshes in the outer Coastal Plain.

Taxonomic Note
Three species occur in North Carolina: *Salicornia ambigua* (chickenclaws), *S. bigelovii* (dwarf saltwort), and *S. virginica* (Virginia glasswort). All are fairly common. Some resources have included the Chenopodiaceae family into Amaranthaceae, but since each family is monophyletic, Weakley (2020) sees no reason to combine them.

Saururus cernuus — Lizard's Tail

Distinctive, drooping flowering spike

Jointed zig-zagging stem

Often in dense colonies

Seedheads straighten out

Elongated, heart-shaped leaves

DICOT HERBS

Saururaceae - Lizard's Tail Family

Saururus cernuus
Lizard's Tail

National Wetland Plant List
Mtn/Pdmt: **OBL**   CP: **OBL**

Coefficient of Conservatism
Mtn: 6   Pdmt: 6   CP: 6

Drawing: Karen Kendig

Habit
Perennial herb to about 0.5 m in height, forming dense stands.

Leaves
Alternate, heart-shaped leaves growing on zig-zagged stems. Leaf venation palmate and leaf edges entire.

Flowers/Fruit
White, graceful spike, which resembles a lizard's tail. Blooms May through September; fruits August to November. However, this plant also reproduces by underground rhizomes and often forms dense stands.

Field Characteristics
Cordate leaves, distinctive white, drooping flowers.

Habitat/Range
Swamps, wet woods and ditches throughout the Coastal Plain and Piedmont.

Similar Species
Pontederia cordata (pickerelweed) has similar leaf bases, but they have unpointed tips and less branched venation. Leaves also rise from base of plant in *P. cordata*.

Solidago patula — Roundleaf Goldenrod

Perennial, 1 to 2 m tall, with erect flowering stem

Large, wide leaves decrease in size along stem

Small, yellow flowers

Widely branching panicles

Serrated basal leaves have winged petioles

DICOT HERBS

Asteraceae - Aster Family

Solidago patula
Roundleaf Goldenrod

National Wetland Plant List
Mtn/Pdmt: **OBL** CP: **OBL**

Coefficient of Conservatism
Mtn: **9** Pdmt: **n/a** CP: **n/a**

Habit
Medium or tall height perennial, 1 to 2 m tall, with basal leaves and erect flowering stem.

Leaves
Alternate, simple, leaves with serrated edges. Basal leaves 15 to 35 cm long. Leaves at base of stem with winged petioles and much larger than those at midstem. Upper leaf surfaces very rough, lower surfaces smooth.

Flowers/Fruit
Small, yellow flowers in open panicle, about as broad as long. Stem below inflorescence is smooth. Flowers and fruits August to early October.

Field Characteristics
Look for basal leaves larger than stem leaves, very rough above and smooth below. Fairly restricted to wetlands.

Habitat/Range
Mountain bogs, seeps, swampy woods, stream banks; only in the Mountains.

Solidago sempervirens — Seaside Goldenrod

Late-branching, leafy plant

Lower leaves in basal rosette

Many small, yellow flowers, mainly on one side of stem

Windborne seeds with tufts of white hairs

Fleshy leaves, brown stems

DICOT HERBS

Asteraceae - Aster Family

National Wetland Plant List
Mtn/Pdmt: **n/a** CP: **FACW**

Coefficient of Conservatism
Mtn: **n/a** Pdmt: **n/a** CP: **6**

Solidago sempervirens
Seaside Goldenrod

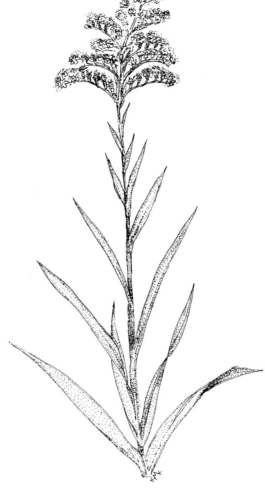

Drawing: Karen Kendig

Habit
Medium to tall (0.4 to 2 m) perennial herb with erect flowering stems arising from rosetted basal leaves.

Leaves
Alternate, simple, smooth-margined, fleshy lance-shaped leaves; at base, 10 to 40 cm long, with those on stem decreasing in size to the top.

Flowers/Fruit
Small, yellow flowers in narrow or broad panicles at top of plant. Flowers and fruits late August to mid-November.

Field Characteristics
Fleshy leaves, untoothed leaf margins, and relatively large leaves on stems distinguish this goldenrod.

Habitat/Range
Along the coast in salt and brackish marshes, tidal fresh marshes, estuarine shores, interdunal swales, beaches, damp roadsides.

Taxonomic Note
Most of this species in North Carolina is *Solidago sempervirens* var. *mexicana*, which some taxonomists now treat as a separate species, *S. mexicana*.

Symphyotrichum pilosum — Frost Aster

Flowers tend to stem from one side of branch

Hairs on stems and leaf undersides

Bracts beneath flowers have pointed tips

Oblanceolate basal leaves

Flowers are small with 20 to 30 petals

DICOT HERBS

Asteraceae - Aster Family

Symphyotrichum pilosum
Frost Aster

National Wetland Plant List
Mtn/Pdmt: **FAC**   CP: **FAC**

Coefficient of Conservatism
Mtn: **3**   Pdmt: **3**   CP: **3**

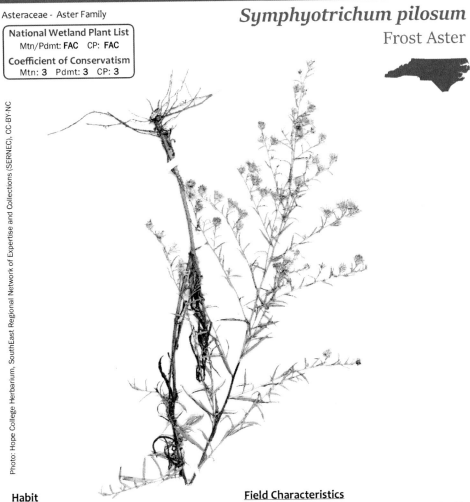

Photo: Hope College Herbarium, SouthEast Regional Network of Expertise and Collections (SERNEC), CC-BY-NC

Habit
Weedy, medium height, erect, perennial aster with white flowers; grows to 1.5 m.

Leaves
Alternate, smooth-margined, green above, gray below. Basal leaves oblanceolate, 2.5 to 10 cm long and wither early. Long spreading hairs on stems and leaf undersides.

Flowers/Fruit
Small, daisy-like flowers, 1.25 to 2 cm wide, with 20 to 30 narrow white petals and yellow or reddish centers; white petals curling under on older flowers. Bracts on flower undersides have green midribs and pointed tips. Flowers typically arranged all on one side of a branch. Fruits are small dry seeds with clusters of hairs for wind-borne dispersal. Flowers and fruits September to November.

Field Characteristics
Plant often leaning. Hairy stems and leaf undersides.

Habitat/Range
As the name implies, this weedy aster is found in old fields, disturbed areas, fencerows, woodland edges, lake and pond shores, dry or moist conditions.

Similar Species
Easily confused with other *Symphyotrichum* species, especially *S. dumosum* (rice button aster), which has flower bracts without sharp-pointed tips and 14 to 20 ray flowers (petals).

Taxonomic Note
Synonym: *Aster pilosus*

Verbena urticifolia — White Vervain

Erect herb with stringy inflorescence

Tiny white flowers on long, thin spikes

Hairy stems and leaves

Dry fruits split into four parts when ripe

Toothed leaves with white center stripe

DICOT HERBS

Verbenaceae - Verbena Family

Verbena urticifolia
White Vervain

National Wetland Plant List
Mtn/Pdmt: **FAC**   CP: **FAC**

Coefficient of Conservatism
Mtn: 3   Pdmt: 3   CP: 3

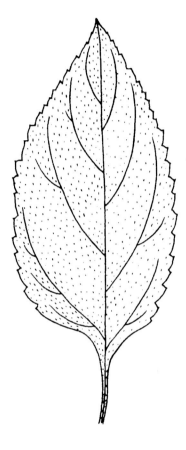

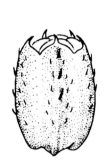

Habit
Very tall, erect herb, 1 to 2.5m tall, with spreading branches.

Leaves
Opposite, lance-shaped, ovate, or oblong leaves with well serrated margins and short petioles. Hairy stems and leaves, with white stripe down center veins of leaves.

Flowers/Fruit
Loose, slender, stringy spikes of tiny, white flowers (3 mm across), on ends of stems and branches. Dry fruits much larger than flowers and split into 4 parts when ripe. Blooms May to November; fruits soon after flowering.

Field Characteristics
Only a few flowers open at a time on each long spike. Stems hairy and four-sided.

Habitat/Range
Grows in a wide variety of moist habitats, such as partly shaded stream banks, floodplain forests, damp swales, fencerows, old fields, meadows. Common throughout, but uncommon near the coast.

VINES

Scientific Name	Common Name	National Wetland Plant List Status Mtns & Piedmont/Coastal Plain	Page
Apios americana	Groundnut	FACW/FACW	358
Berchemia scandens	Alabama Supplejack	FACW/FAC	360
Bignonia capreolata	Crossvine	FAC/FAC	362
Campsis radicans	Trumpet Creeper	FAC/FAC	364
Gelsemium sempervirens	Carolina Jessamine	FAC/FAC	366
Mikania scandens	Climbing Hempvine	FACW/FACW	368
Nekemias arborea	Peppervine	FACW/FAC	370
Smilax bona-nox	Saw Greenbrier	FACU/FAC	372
Smilax glauca	Whiteleaf Greenbrier	FACU/FAC	374
Smilax laurifolia	Laurel Greenbrier	OBL/FACW	376
Smilax rotundifolia	Roundleaf Greenbrier	FAC/FAC	378
Toxicodendron radicans	Eastern Poison Ivy	FAC/FAC	380
Vitis cinerea	Graybark Grape	FACW/FAC	382

Apios americana — Groundnut

Vine with 5 to 7 leaflets

"Bean" flowers connect to center stalk in inflorescence

Pink, lavender, or brownish flowers

Fruits resemble green beans, drying to light brown

Often sprawling over other wetland plants

VINES

Fabaceae - Legume Family

National Wetland Plant List
Mtn/Pdmt: **FACW**   CP: **FACW**

Coefficient of Conservatism
Mtn: 5   Pdmt: 5   CP: 5

Apios americana
Groundnut

Habit
Twining herbaceous vine, spreading by rhizomes.

Leaves
Compound leaves with 5-7 leaflets, 3-6 cm long with pointed tips. Leaves 10-20 cm long.

Flowers/Fruit
Typical "bean" flower, purplish to brown. Flower roughly 2- lipped with 5 petals bearing a long, bean-like pod, up to 10 cm. Blooms June to August; fruits July to September.

Field Characteristics
Recognizable in wetlands by its compound leaves with 5 to 7 leaflets. Herbaceous vine with typical "bean family" characteristics (flower, pea-pod). Underground tuber is edible.

Habitat/Range
Freshwater marshes, edges of streams or ponds, bottomlands throughout NC.

Berchemia scandens — Alabama Supplejack

Smooth thick vines

Loose hanging vine

Prominent, straight veins in leaves

Green fruits mature to blue-black

Smooth, reddish brown stems

VINES

Rhamnaceae - Buckthorn Family

National Wetland Plant List
Mtn/Pdmt: **FACW**   CP: **FAC**

Coefficient of Conservatism
Mtn: **n/a**   Pdmt: **6**   CP: **6**

Berchemia scandens
Alabama Supplejack

Drawing: Mark Mohlenbrock, USDA Wetland Flora Field Office Guide

Habit
Deciduous, climbing, flexible, woody vine.

Leaves
Alternate, ovate or elliptic, shiny, entire leaves, 4-8 cm long and 3 cm wide. Leaf venation strikingly parallel with 10 or more straight veins on each half of the leaf.

Flowers/Fruit
Small, inconspicuous flowers in panicles. Fruit an elliptical dark blue or black drupe, 5-7 mm long. Blooms April and May; fruits August to October.

Field Characteristics
Smooth, reddish-brown stems, useful in basketry. May be toxic; related to other toxic plants.

Habitat/Range
Floodplain forests, moist sandy woods, stream banks, flat woods, bottomlands, rich woodlands, mainly in the Coastal Plain.

Bignonia capreolata — Crossvine

Thin, smooth, climbing vine

Curved leaf base

Curved tubular red and yellow flowers

Fruits are long bean-like capsules

Green or reddish fused sepals remain after petals fall and fruits start to grow (inset)

VINES

Bignoniaceae - Bignonia Family

National Wetland Plant List
Mtn/Pdmt: **FAC**   CP: **FAC**

Coefficient of Conservatism
Mtn: **5**   Pdmt: **5**   CP: **4**

Bignonia capreolata
Crossvine

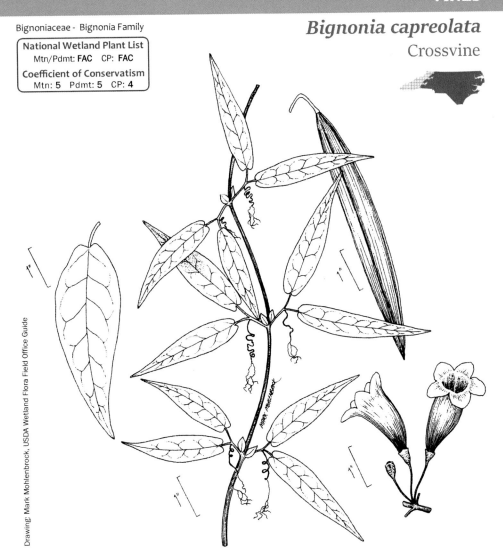

Drawing: Mark Mohlenbrock, USDA Wetland Flora Field Office Guide

Habit
Climbing woody semi-evergreen vine, which can reach great heights, climbing by branched tendrils.

Leaves
Pairs of opposite leaflets, oblong or elongate-cordate up to 15 cm long and 2-7 cm wide.

Flowers/Fruit
Showy, yellow and orange/red tubular flowers in early spring, forming fruit which is a long flattened bean-like capsule up to 15 cm long. Blooms April/May; fruits July/August.

Field Characteristics
Distinctive paired leaves. When severed, the large vines of this species exhibit a large cross as the name implies.

Habitat/Range
Swamps, moist woods, bottomlands and bay forests; chiefly Coastal Plain and Piedmont, infrequent in the mountains.

Similar Species
Flowers similar to *Campsis radicans* (trumpet creeper) but *Bignonia capreolata* (crossvine) flowers are more yellow on petals and bloom earlier.

Taxonomic Note
Synonym: *Anisostichus capreolata*

Campsis radicans — Trumpet Creeper

Red-orange, tubular flowers appear in summer

Flaky, light-colored vine with holdfasts

Compound leaves with 7 to 11 toothed leaflets

High-climbing vine, becomes folded/ridged with age

Mature fruits are brown capsules with many winged seeds

VINES

Bignoniaceae - Bignonia Family

National Wetland Plant List
Mtn/Pdmt: **FAC**   CP: **FAC**

Coefficient of Conservatism
Mtn: **2**   Pdmt: **2**   CP: **2**

Campsis radicans
Trumpet Creeper

Toxic Plant

Drawing: Karen Kendig

Habit
Deciduous, climbing or sprawling trailing woody vine. Trumpet creeper climbs via aerial roots (holdfasts) in double rows on stems and does not have tendrils.

Leaves
Opposite, pinnately compound leaves with 7-11 toothed leaflets, usually up to 4-8 cm long. Compound leaves up to 30 cm long.

Flowers/Fruit
Terminal cluster of 2-9 showy red "trumpet" flowers with 5 lobes at the end of the tube later forming brown capsule with winged seeds. Blooms June/July; fruits September/October.

Field Characteristics
Vine is flaky, tan-colored, with holdfasts, becoming ridged or folded inward with age. Elongated capsule fruits with winged seeds are distinctive. Toxic if eaten.

Habitat/Range
Forested wetlands, moist uplands, old fields, fence rows, waste places, throughout NC.

Similar Species
Leaves similar to *Nekemias arborea* (peppervine), but *Campsis radicans* (trumpet creeper) produces seed capsules instead of berries. See Common Confusions section, p. 409. Leaves also similar to *Wisteria* spp. which have wavy-margined leaves and vines with no holdfasts. Flowers similar to *Bignonia capreolata* (crossvine) but are more red and bloom later.

Gelsemium sempervirens — Carolina Jessamine

Showy, yellow flowers on new growth

Leaf shape varies little

Blooms February to early May

Capsule fruit

Dry, elongated capsules split when mature.

VINES

Loganiaceae - Logan Family

National Wetland Plant List
Mtn/Pdmt: **FAC** CP: **FAC**

Coefficient of Conservatism
Mtn: **n/a** Pdmt: **4** CP: **4**

Gelsemium sempervirens
Carolina Jessamine

Highly Toxic Plant

Drawing: Mark Mohlenbrock, USDA Wetland Flora Field Office Guide

Habit
Slender woody evergreen vine, to 3 m long, sometimes climbing to treetops.

Leaves
Opposite, lanceolate shiny leaves with entire margins.

Flowers/Fruit
Clusters of showy yellow flowers, on new growth, blooming February to early May. Fruits are elongated, splitting capsules. Blooms February to early May; fruits September to November.

Field Characteristics
Leaf shape varies little. Entire plant is toxic.

Habitat/Range
Occurs both in moist and dry areas, in hardwood and pine forests, along fencerows, and in bottomlands.

Similar Species
When trailing along the ground, could be confused with *Lonicera japonica* (Japanese honeysuckle), but *G. sempervirens* (Carolina jessamine) leaves are shiny, narrower, and more elongated.

Mikania scandens — Climbing Hempvine

Heart or triangular-shaped leaves; stems 4-sided

Flowers profusely in sunny, wet areas

Fruits are nutlets in clusters, with whitish barbed bristles

Small white flowers in clusters

Sprawling, climbing herbaceous vine

VINES

Asteraceae - Aster Family

National Wetland Plant List
Mtn/Pdmt: **FACW**   CP: **FACW**

Coefficient of Conservatism
Mtn: **n/a**   Pdmt: **4**   CP: **4**

Mikania scandens
Climbing Hempvine

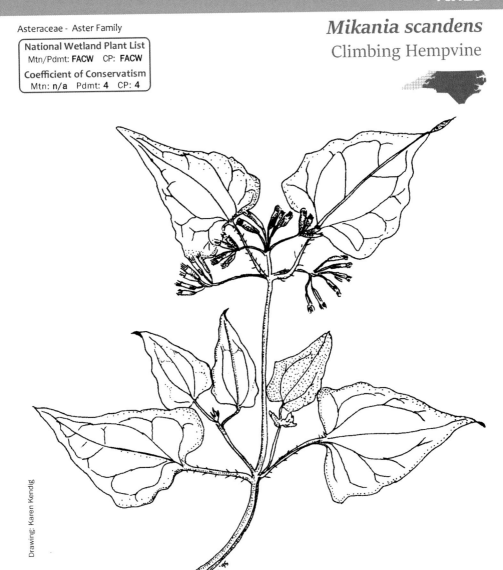

Drawing: Karen Kendig

Habit
Deciduous, climbing, herbaceous vine, often blanketing nearby vegetation.

Leaves
Opposite, cordate or triangular leaves with long petioles. Leaves 3-13 cm long with 3 main distinctive veins.

Flowers/Fruit
White flowers in heads originating in axils of leaves. Fruits are nutlets in clusters, with whitish barbed bristles. Flowering and fruiting July to October.

Field Characteristics
Attractive, aggressive herbaceous vine which climbs clockwise and upward. Stems 4-sided.

Habitat/Range
Sunny perimeters of lakes, swamps, wet woodlands, freshwater marshes, stream banks in the Coastal Plain and Piedmont.

Nekemias arborea — Peppervine

Twice divided compound leaves

Berries are poisonous

Can be weedy

Purplish stems

Greenish-white flowers in clusters, June to October

VINES

Vitaceae - Grape Family

Nekemias arborea
Peppervine

National Wetland Plant List
Mtn/Pdmt: **FACW** CP: **FAC**

Coefficient of Conservatism
Mtn: **n/a** Pdmt: **4** CP: **4**

Toxic Fruit

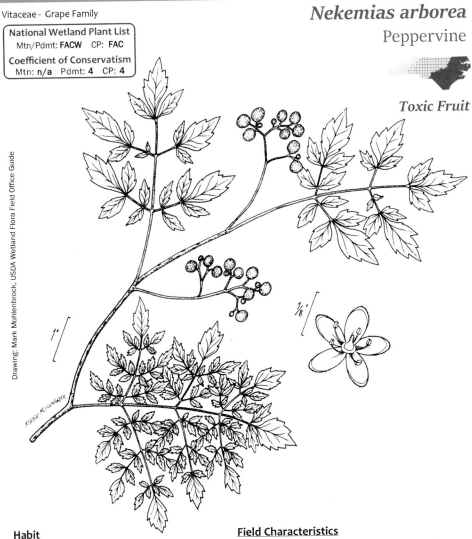

Drawing: Mark Mohlenbrock, USDA Wetland Flora Field Office Guide

Habit
Deciduous, climbing woody vine with few or no tendrils; sometimes bush-like or low growing.

Leaves
Alternate, bi-pinnately divided and up to 6 inches long and wide. 1-3 pairs of leaflets ovate and coarsely toothed.

Flowers/Fruit
Inconspicuous greenish-white flowers in flat-topped clusters. Poisonous berries mature from green to pink to blue-black. Flowers June to October, fruiting soon after.

Field Characteristics
Look for twice divided compound leaves. Berries poisonous and look similar to grapes.

Habitat/Range
Found predominantly in the Coastal Plain in a wide variety of moist to wet sites, wetlands and stream banks. Can be weedy, but rarely found in acidic wetlands (pocosins, bays, Sandhill streams).

Similar Species
Leaves similar to *Campsis radicans* (trumpet creeper), but *Nekemias arborea* (peppervine) has alternate leaves, double pinnately compound. See Common Confusions section, p. 409.

Taxonomic Note
Synonyms: *Ampelopsis arborea*; *Ampelopsis bipinnata*

Smilax bona-nox — Saw Greenbrier

Flower stalks longer than leaf petioles

Black grape-like berries, September to November

Usually cordate leaf shape; mottled green

Somewhat variable leaf shape

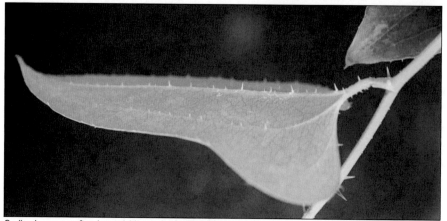
Smilax bona-nox often has prickles on leaf underside

VINES

Smilacaceae - Catbrier Family

Smilax bona-nox
Saw Greenbrier

National Wetland Plant List
Mtn/Pdmt: **FACU**   CP: **FAC**

Coefficient of Conservatism
Mtn: **4**   Pdmt: **4**   CP: **4**

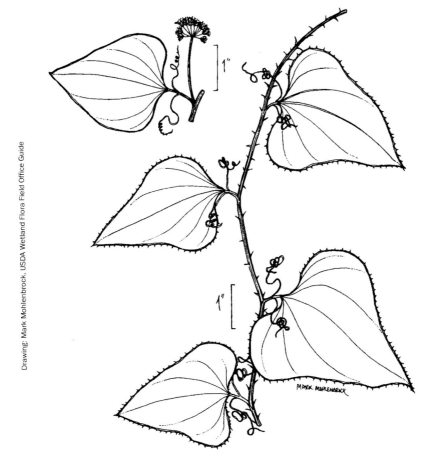

Habit
Semi-evergreen, high climbing, thorny woody vine with paired tendrils. Stem is slightly square.

Leaves
Alternate, leathery cordate leaves usually with small spines around leaf edges. Mottled green leaves 10 cm long and 8 cm wide. Underside light green.

Flowers/Fruit
Small flowers in umbels, producing black grape-like berries. Stalks of flowering umbels longer than leaf petioles. Flowers late April to May; fruits September into November.

Field Characteristics
Attractive cordate or hastate leaves, often mottled and rimmed with bristles.

Habitat/Range
Upland woods, floodplains, bay forests, pine flats and in cutover areas. Found throughout the state, but most commonly in the Coastal Plain.

Smilax glauca — Whiteleaf Greenbrier

Stems have whitish coating (glaucous)

Flowering umbels greenish yellow

Leaves whitish on underside

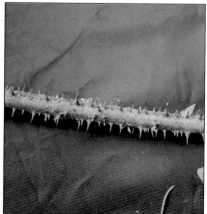
Lower stems often densely thorny

Leaves sometimes mottled on many *Smilax* species

VINES

Smilacaceae - Catbrier Family

National Wetland Plant List
Mtn/Pdmt: **FACU**   CP: **FAC**

Coefficient of Conservatism
Mtn: **4**   Pdmt: **4**   CP: **4**

Smilax glauca
Whiteleaf Greenbrier

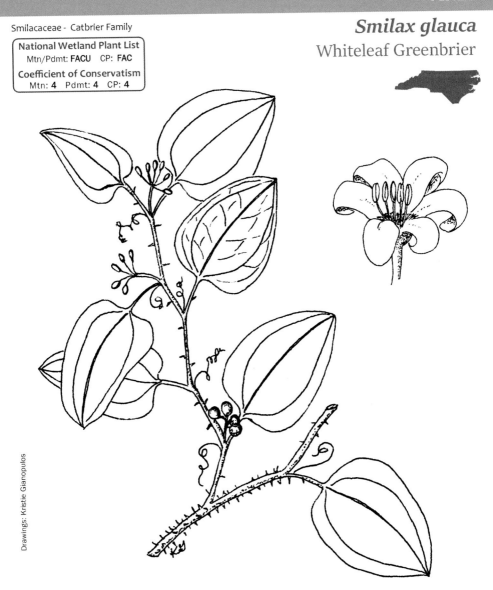

Drawings: Kristie Gianopulos

Habit
High climbing, thicket-forming, thorny vine.

Leaves
Alternate, simple, ovate to lance-shaped, not leathery. Undersides whitish.

Flowers/Fruit
Flowers in umbels with stalks 1.5 to 3 times longer than leaf petioles. Berries shiny black at maturity, persisting through winter. Blooms late April to early June; fruits September to November.

Field Characteristics
Look for leaves with whitish undersides and young stems with a glaucous coating. Spines have dark tips and lower stems often densely thorny.

Habitat/Range
In both drier and mesic areas: fence rows, old fields, woodlands, floodplain forests, pocosins, swamps.

Smilax laurifolia — Laurel Greenbrier

Leathery, linear leaves, with pointed tips

Vines can be large with stout prickles

Leaves often upturned; small white or light green flowers

Fruits on short stalks; black at maturity

Thick, leathery leaves with prominent midribs on undersides

VINES

Smilacaceae - Catbrier Family

Smilax laurifolia
Laurel Greenbrier

National Wetland Plant List
Mtn/Pdmt: **OBL**   CP: **FACW**

Coefficient of Conservatism
Mtn: **n/a**   Pdmt: **7**   CP: **5**

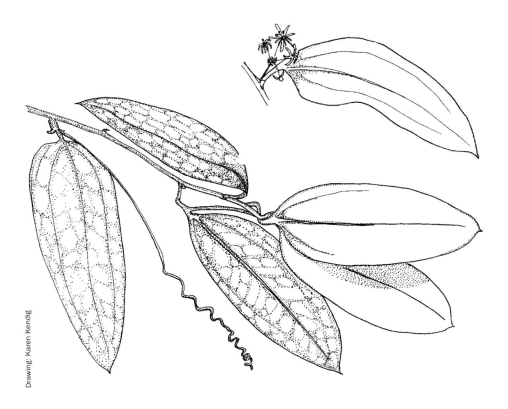

Drawing: Karen Kendig

Habit
Vigorous evergreen, woody vine with thorns, forming dense thickets in wet areas.

Leaves
Alternate, simple, narrowly oblong, thick leathery leaves up to 10-15 cm long and 5 cm wide. Leaves usually pointing upward and may appear mottled.

Flowers/Fruit
Light green flowers in umbels, producing black spherical berries, about 1 cm wide. Flowers July/August; fruits mature September/October of the following year.

Field Characteristics
Very leathery leaves oblong and pointing upward, prominent midrib vein on leaf undersides. Older stems can be very thick.

Habitat/Range
Swamps, bay forests, pine flats, pocosins in the Coastal Plain and in the southeastern Piedmont of NC, often in standing water. Found in a few Mountain counties.

Smilax rotundifolia — Roundleaf Greenbrier

A very common *Smilax* species

Small green fruits mature to black

Leaves can be quite large

Immature fruits

Leaf shape can be triangular or rounded, with no spines on midrib

VINES

Smilacaceae - Catbrier Family

National Wetland Plant List
Mtn/Pdmt: **FAC**   CP: **FAC**

Coefficient of Conservatism
Mtn: **4**   Pdmt: **4**   CP: **4**

Smilax rotundifolia
Roundleaf Greenbrier

Drawing: Mark Mohlenbrock, USDA Wetland Flora Field Office Guide

Habit
High climbing, semi-evergreen, thorny vine with green stems and tendrils.

Leaves
Alternate, simple, shiny, ovate to circular and pointed at the tip, sometimes quite large, with short petioles. Underside green with a spineless midrib. Leaves not very leathery.

Flowers/Fruit
Stalks of umbel with light green flowers not longer than leaf petioles, as in *Smilax bona-nox* (saw greenbrier). Fruits blue-black berries, persisting into winter. Flowers late April/ May; fruits September to November.

Field Characteristics
Underside of leaves green with spineless midrib; wide leaves, relative to many other *Smilax* species.

Habitat/Range
Dry-mesic to mesic forests, bottomland and riparian forests, swamps, pond margins, pine wetlands, old fields, fencerows, roadsides.

Toxicodendron radicans — Eastern Poison Ivy

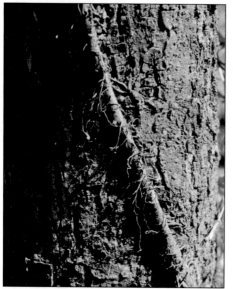

Hairy vine is a key feature

Vines can be large enough to look like part of a tree

Leaves sometimes untoothed

White or light green flowers in panicles

Leaves sometimes have teeth or small lobes

VINES

Anacardiaceae - Cashew or Sumac Family

National Wetland Plant List
Mtn/Pdmt: **FAC**   CP: **FAC**

Coefficient of Conservatism
Mtn: 3   Pdmt: 3   CP: 3

Toxicodendron radicans
Eastern Poison Ivy

Highly Toxic Plant

Drawing: Karen Kendig

Habit
Deciduous, high-climbing, woody vine with distinctive hairy, aerial roots. When unsupported, may take a shrubby growth form of slender, unbranched stems up to 1 m tall.

Leaves
Alternate, shiny, thin leaves with 3 ovate (sometimes toothed) leaflets with pointed ends. Leaves variable but end leaflet is on a long stalk.

Flowers/Fruit
Panicles of white or light green nondescript flowers later produce light green or white, berry-like drupes that often persist into winter. Blooms late April through May; fruits August to October.

Field Characteristics
"Leaves of three, let them be," is a good reminder with this plant, which can cause a severe rash in people allergic to poison ivy. Hairy vines and upright branch tips are helpful in identification. Seeds are poisonous.

Habitat/Range
Swamps, wetlands, dry uplands, throughout NC.

Similar Species
Acer negundo (box elder) (tree) has very similar leaves to *Toxicodendron* radicans (eastern poison ivy) but its growth form is more robust, with green twigs. Saplings may be difficult to distinguish from eastern poison ivy. See Common Confusions section, p. 403. *Parthenocissus quinquefolia* (Virginia creeper) may be mistaken for *T. radicans*, but has sets of five leaves instead of sets of three. Young plants can be confused with *Arisaema triphyllum* (Jack-in-the-pulpit); see Common Confusions section, p. 410.

Vitis cinerea — Graybark Grape

Flowers in long, branching stems

Edible grapes mature September/October

Younger branches angle off stem; older bark flakes

Leaves sometimes deeply lobed

Backs of leaves cobweb hairy

VINES

Vitaceae - Grape Family

Vitis cinerea
Graybark Grape

National Wetland Plant List
Mtn/Pdmt: **FACW**   CP: **FAC**

Coefficient of Conservatism
Mtn: 5   Pdmt: 5   CP: 5

Habit
Deciduous, climbing woody vine with tendrils opposite the leaves, often climbing into the canopies of trees.

Leaves
Alternate, cordate, or 3 to 7 lobed and toothed. Undersides cobweb hairy. Young branches angled off main stem.

Flowers/Fruit
Small flowers in panicles; fruit an edible grape. Flowers June; fruits mature September/October.

Field Characteristics
Climbing vine with cordate or lobed leaves. Older vines have shredded bark and younger twigs are smooth. Sometimes aerial roots extend downward from vines.

Habitat/Range
Low woods, floodplains, stream banks, bottomlands throughout NC.

Similar Species
Vitis (grape) vine is woody and brown, shaggy when mature (except muscadine grapes); *Campsis radicans* (trumpet creeper) is light tan, and *Toxicodendron radicans* (eastern poison ivy) has many, hairy, aerial roots. *Vitis* vines often hang loosely, unlike *Parthenocissus quinquefolia* (Virginia creeper) and *T. radicans*.

Taxonomic Note
Weakley treats *V. cinerea* var. *baileyana* (calling it *V. baileyana*) and *V. cinerea* var. *cinerea* (calling it *V. simpsonii*) as full species. Differences lie in pubescence of twigs and node pigmentation.

AQUATICS

Scientific Name	Common Name	National Wetland Plant List Status Mtns & Piedmont/Coastal Plain	Page
Azolla caroliniana	Carolina Mosquitofern	OBL/OBL	386
Lemna spp.	Duckweed	OBL/OBL	388
Myriophyllum spp.	Water-milfoil	OBL/OBL	390
Nelumbo lutea	American Lotus	OBL/OBL	392
Nuphar lutea	Yellow Pond-Lily	OBL/OBL	394
Nymphaea odorata	American Water-lily	OBL/OBL	396
Utricularia spp.	Bladderwort	OBL/OBL	398

Azolla caroliniana — Carolina mosquitofern

Reddish, overlapping, scale-like leaves

Can blanket the surface of sunny, stagnant water

Appears clumped into small clusters on water

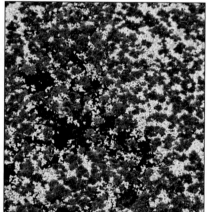
Often mixed with duckweed

Small plant; more reddish in direct sun

AQUATICS

Azollaceae - Mosquitofern Family

National Wetland Plant List
Mtn/Pdmt: **OBL** CP: **OBL**

Coefficient of Conservatism
Mtn: **n/a** Pdmt: **3** CP: **3**

Azolla caroliniana
Carolina mosquitofern

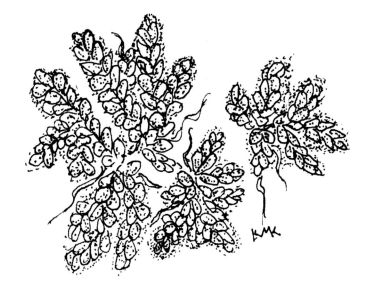

Drawing: top - Patrick Guénette; bottom - Karen Kendig

Habit
Small, free-floating, red-tinged, aquatic plant, with a similar habit to duckweed.

Leaves
Green to red, resembling branching, overlapping scales.

Flowers/Fruit
Inconspicuous.

Field Characteristics
Unlike *Lemna* spp. (duckweed), this tiny, floating, aquatic plant has a reddish hue and scaled appearance.

Habitat/Range
Beaver ponds, floodplain backwaters, stagnant waters of old millponds; chiefly in the Coastal Plain.

Lemna spp. — Duckweed

Usually found crowded on water surface

Lemna spp. have one root per frond.

Each leaf (frond) has a ridge down the middle

Grows in extensive colonies, covering large areas of sluggish creeks, swamps, and ponds

AQUATICS

Araceae - Arum Family

National Wetland Plant List
Mtn/Pdmt: **OBL**   CP: **OBL**
Coefficient of Conservatism
Varies

Lemna spp.
Duckweed

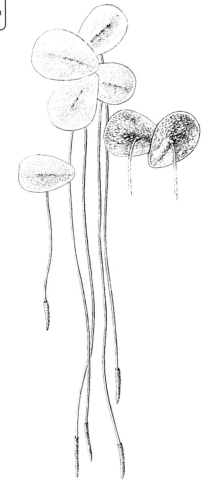

Drawing: Courtesy of the Flora of North America Association, John Myers

Habit
Small, flattened, floating, aquatic plant occurring on the water's surface.

Leaves
Leaves and stems reduced and referred to as 'fronds' which look like small leaves, 2-6 mm long. Fronds of *Lemna* light green with a middle ridge, containing one root per frond.

Flowers/Fruit
Inconspicuous flowers produced in a pouch or spathe, but this plant seldom flowers.

Field Characteristics
Tiny, light green leaves (fronds) have 1 root per frond.

Habitat/Range
Swamps, ponds, lakes, and sluggish creeks in the Coastal Plain and Piedmont. Occasionally in the Mountains.

Similar Species
The single root per frond helps to distinguish *Lemna* species from similar types of duckweed. Five species of *Lemna* occur in North Carolina.

Taxonomic Note
The family Araceae is the currently accepted name, with a family synonym of Lemnaceae.

Myriophyllum spp.   Water-milfoil

Leaves whorled in sets of 3 to 6

Myriophyllum aquaticum (non-native)

Plant may be partly submerged (*M. heterophyllum* shown)

Leaves finely toothed, feather-like

Myriophyllum spp. have long trailing stems and finely dissected leaves. Compare to *Proserpinaca* spp. (inset).

AQUATICS

Haloragaceae - Water Milfoils Family

National Wetland Plant List
Mtn/Pdmt: **OBL**   CP: **OBL**

Coefficient of Conservatism
Mtn: **8**   Pdmt: **5**   CP: **5**

Myriophyllum spp.
Water-milfoil

One species non-native
(Myriophyllum aquaticum)

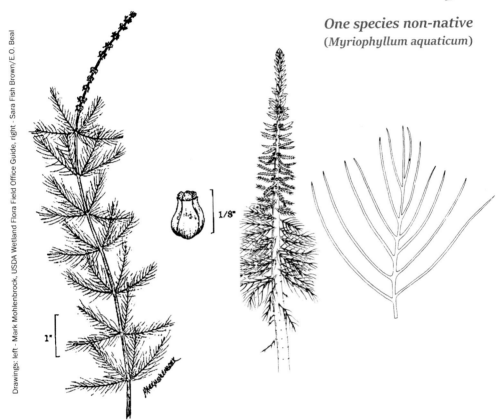

Habit
Feathery aquatic plant with long trailing stems.

Leaves
Finely toothed, whorled in sets of 3 to 6; may be fully submerged or emerging from the surface.

Flowers/Fruit
Inconspicuous flowers at bases of leaves, but usually not flowering at all. Flowers and fruits April to September/October.

Field Characteristics
Leaves feather-like and whorled; often not flowering.

Habitat/Range
Ponds, lakes, ditches, sluggish streams.

Similar Species
Can be confused with the less common *Proserpinaca* spp., which have similar, but flattened and much less feathery, leaves (see photo comparison, left).

Taxonomic Note
The most common species in North Carolina are *Myriophyllum aquaticum* (parrot feather water-milfoil) and *M. heterophyllum* (twoleaf water-milfoil), both of which are obligate wetland species. *M. aquaticum* is a non-native with light green, feathery leaves that rise above the water surface. *M. heterophyllum* is a native, mostly Coastal Plain, species, with brownish-green leaves seldom above the surface, except the thick flowering stalks, which have undivided, lance shaped leaves.

Nelumbo lutea — American Lotus

Large flowers and circular leaves

Can form colonies

Large, pointed buds; flat-topped fruit

Distinctive seedhead

Beautiful, unique flower

AQUATICS

Nelumbonaceae - Water lotus Family

National Wetland Plant List
Mtn/Pdmt: **OBL**   CP: **OBL**
Coefficient of Conservatism
Mtn: **n/a**   Pdmt: **5**   CP: **5**

Nelumbo lutea
American Lotus

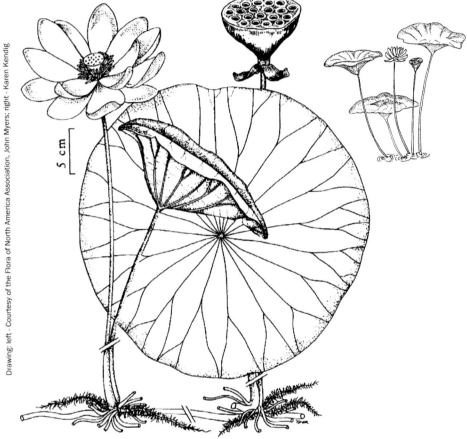

Drawing: left - Courtesy of the Flora of North America Association, John Myers; right - Karen Kendig

Habit
Immersed aquatic plant with large, round leaves which extend above water surface. Spongy rhizomes from which leaves and flowers arise.

Leaves
Large, round, bluish-green leaves, 20-70 cm wide with concave centers. Early leaves float on surface, later becoming suspended above water as petioles continue to grow.

Flowers/Fruit
Attractive, pale yellow flower can be as large as the leaves. Seed pods woody and persistent; often used ornamentally. Blooms June to September; fruits late summer/early fall.

Field Characteristics
Large leaves of this plant not split as with water lilies. Distinctive seed pods persist. A milky substance is found in all parts of American lotus.

Habitat/Range
Ponds, lakes, marshes, and slow streams.

Similar Species
Can be confused with *Nymphaea odorata* (American water-lily), but *Nelumbo lutea* (American lotus) leaves are unsplit and flowers are pale yellow, not white.

Nuphar lutea — Yellow Pond-Lily

Unbranched venation in leaves

Rounded, arrow-head leaf shape is distinctive

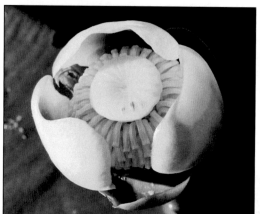
Yellow flowers appear to barely open

Fruits are green and yellow

Can be locally abundant; leaves may be elevated above water surface

AQUATICS

Nymphaeaceae - Water lily Family

Nuphar lutea
Yellow Pond-Lily

National Wetland Plant List
Mtn/Pdmt: **OBL**   CP: **OBL**

Coefficient of Conservatism
Mtn: **6**   Pdmt: **5**   CP: **5**

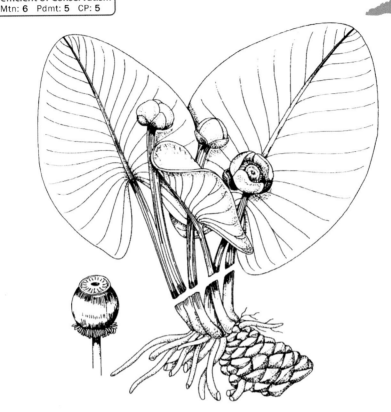

Drawing: Courtesy of the Flora of North America Association, John Myers

Habit
Immersed, floating or submersed aquatic plant growing from a spongy rhizome.

Leaves
Leaves ovate or cordate with split leaf bases, about 30 cm long and 25 cm wide. Veins originate from main central vein and extend to outer edge of leaf. Leaves may be elevated above water surface.

Flowers/Fruit
Spherical yellow flowers with 6 light green sepals and numerous petals. Flowers and fruits April to October.

Field Characteristics
Cordate leaves with prominent midvein and parallel side veins, and distinctive spherical yellow flowers. In flowing coastal waters, leaves 3 times as long as wide and margins undulate or rippled. Intermediates with varying length to width ratios also occur.

Habitat/Range
Lakes, ponds, swamps and streams throughout the state, but predominantly in the southern Coastal Plain.

Similar Species
Can be confused with *Nymphaea odorata* (American water-lily), but *Nuphar lutea* (yellow pond-lily) leaves are more rounded and flower is yellow. See Common Confusions section, p. 410.

Taxonomic Note
Nuphar lutea ssp. *advena* is the common subspecies in North Carolina; synonym: *Nuphar advena*.

Nymphaea odorata — American Water-lily

Showy, fragrant flower; sometimes pink-tinged

Leaves always floating on water surface

Flowers sometimes green-tinged

Sometimes in extensive colonies in waterbodies

Branched venation in leaves; pointed tips at split; purplish undersides.

AQUATICS

Nymphaeaceae - Water lily Family

Nymphaea odorata
American Water-lily

National Wetland Plant List
Mtn/Pdmt: **OBL**   CP: **OBL**

Coefficient of Conservatism
Mtn: **6**   Pdmt: **4**   CP: **5**

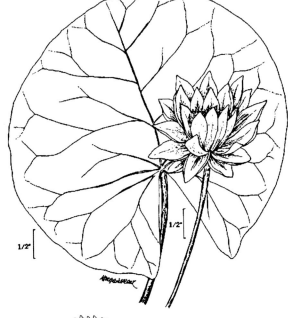

Habit
Floating-leaved, aquatic plant with long leaf stalks arising from the rhizome.

Leaves
Circular leaves, split with pointed lobes. Leaves green on surface and purplish on undersides. Leaves to about 25 cm wide and long. Veins branching.

Flowers/Fruit
White or pinkish fragrant flower with 25 or more ovate petals. Flower has 4 green sepals. Flowers and fruits June to September.

Field Characteristics
Split leaf with reddish undersides and fragrant flower when in bloom.

Habitat/Range
Lakes, ponds, and slow streams throughout the state.

Similar Species
The similar *Nuphar lutea* (yellow pond-lily) leaf has a rounded or cordate base and green underside. *N. lutea* leaves may become elevated above the water surface; this is not the case with *Nymphaea odorata* (American water-lily). See Common Confusions section, p. 410.

397

Utricularia spp. — Bladderwort

Sometimes forms floating mats

Unique yellow, white, or purple flowers

Flowering stems rise well above water surface

Purple less common (*Utricularia purpurea*)

Bladders capture tiny aquatic animals by activating special trigger hairs to trap them

AQUATICS

Lentibulariaceae - Bladderwort Family

Utricularia spp.
Bladderwort

National Wetland Plant List
Mtn/Pdmt: **OBL** CP: **OBL**
Coefficient of Conservatism
Varies

Drawing: Bodor Tivadar

Habit
Rootless, herbaceous, aquatic plant (though a few are terrestrial) containing floating bladders, sometimes forming mats. Stems thin and dendritically branched.

Leaves
Inconspicuous linear leaves, usually alternate, but may be whorled or opposite.

Flowers/Fruit
Long, flowering stem with 1-20 flowers, usually yellow, occasionally white or purple, and rising above water surface. Flower two-lipped with upper lip 2-lobed and lower lip 3-lobed. Flowers May or June to September/October; fruits soon after flowering.

Field Characteristics
Distinctive bladders and thinly dissected leaves help this plant to float upright for photosynthesis. Small bladders also trap and consume tiny aquatic invertebrates for nourishment.

Habitat/Range
Aquatic plants floating in water of ponds, lakes, shallow pools. Some species found statewide but the majority are restricted to the Coastal Plain.

Similar Species
Fourteen species occur in North Carolina, none of which are very widespread. The most widespread species in North Carolina are *Utricularia juncea* (southern bladderwort) and *U. subulata* (zigzag bladderwort). The lavender *U. purpurea* (eastern purple bladderwort) is less common than yellow species.

COMMON CONFUSIONS
Illustrated Comparisons

Table of Contents for Comparisons
Scientific Name (Common Name) — **Page**

TREES
Betula spp. (Birches) .... versus .... *Carpinus caroliniana* (Ironwood) 402
Betula spp. (Birches) .... versus .... *Ulmus americana* (American Elm) 402
Acer negundo (Boxelder) .... versus .... *Toxicodendron radicans* (E. Poison Ivy) 403

SHRUBS
Baccharis halimifolia (Eastern Baccharis) .... versus .... *Iva frutescens* (Marsh Elder) 403
Clethra alnifolia (Coastal Sweet-pepperbush) .... versus .... *Itea virginica* (Virginia Sweetspire) 404
Eubotrys racemosus (Swamp Fetterbush) .... versus .... *Itea virginica* (Virginia Sweetspire) 404
Aesculus sylvatica (Painted Buckeye) .... versus .... *Asimina triloba* (Common Pawpaw) 405

FERNS
Onoclea sensibilis (Sensitive Fern) .... versus .... *Woodwardia areolata* (Netted Chain Fern) 405
Osmundastrum cinnamomeum (Cinn. Fern) .... versus .... *Woodwardia virginica* (Virginia Chain Fern) 406

MONOCOT HERB
Iris spp. (Iris) (Emerging) .... versus .... *Typha latifolia* (Broadleaf Cattail) (Emerging) . 406
Pontedaria cordata (Pickerelweed) .... versus .... *Sagittaria latifolia* (Broadleaf Arrowhead) 407
Peltandra virginica (Green Arrow Arum) .... versus .... *Sagittaria latifolia* (Broadleaf Arrowhead) 407
Phragmites australis (Common Reed) .... versus .... *Spartina cynosuroides* (Big cordgrass) 408

DICOT HERBS
Boehmeria cylindrica (False Nettle) .... versus .... *Pilea pumila* (Canadian Clearweed) 408
Centella erecta (Erect Centella) .... versus .... *Hydrocotyle umbellata* (Marsh Pennywort) 409

VINES
Campsis radicans (Trumpet Creeper) .... versus .... *Nekemias arborea* (Peppervine) 409
Toxicodendron radicans (E. Poison Ivy) .... versus .... *Arisaema triphyllum* (Jack-in-the-Pulpit) 410

AQUATICS
Nuphar lutea (Yellow Pond-Lily) .... versus .... *Nymphaea odorata* (American Water-lily) 410

Index of Compared Species

Scientific Name (Common Name)	Page	Scientific Name (Common Name)	Page
Acer negundo (Boxelder)	403	*Nekemias arborea* (Peppervine)	409
Aesculus sylvatica (Painted Buckeye)	405	*Nuphar lutea* (Yellow Pond-Lily)	410
Arisaema triphyllum (Jack-in-the-Pulpit)	410	*Nymphaea odorata* (American Water-lily)	410
Asimina triloba (Common Pawpaw)	405	*Onoclea sensibilis* (Sensitive Fern)	405
Baccharis halimifolia (Eastern Baccharis)	403	*Osmundastrum cinnamomeum* (Cinn. Fern)	406
Betula spp. (Birches)	402	*Peltandra virginica* (Green Arrow Arum)	407
Boehmeria cylindrica (False Nettle)	408	*Phragmites australis* (Common Reed)	408
Campsis radicans (Trumpet Creeper)	409	*Pilea pumila* (Canadian Clearweed)	408
Carpinus caroliniana (Ironwood)	402	*Pontedaria cordata* (Pickerelweed)	407
Centella erecta (Erect Centella)	408	*Sagittaria latifolia* (Broadleaf Arrowhead)	407
Clethra alnifolia (Coastal Sweet-pepperbush) .	404	*Spartina cynosuroides* (Big cordgrass)	408
Eubotrys racemosus (Swamp Fetterbush)	404	*Toxicodendron radicans* (E. Poison Ivy)	403,410
Hydrocotyle umbellata (Marsh Pennywort)	409	*Typha latifolia* (Broadleaf Cattail) (Emerging) ...	406
Iris spp. (Iris) (Emerging)	406	*Ulmus americana* (American Elm)	402
Itea virginica (Virginia Sweetspire)	404	*Woodwardia areolata* (Netted Chain Fern)	405
Iva frutescens (Marsh Elder)	403	*Woodwardia virginica* (Virginia Chain Fern)	406

Illustrated Comparisons

***Betula* spp.**
Birches

Look for: Peeling bark; symmetrical, triangular leaf bases. p 27.

Carpinus caroliniana
Ironwood

Look for: Muscular trunk; symmetrical, rounded leaf bases. p 29.

• •

***Betula* spp.**
Birches

Look for: Leaf base symmetrical and triangular; doubly serrated leaf margins. Peeling bark. p 27.

Ulmus americana
American Elm

Look for: Leaf base asymmetrical; larger leaves. Dark, furrowed bark; trunk often buttressed. p 85.

COMMON CONFUSIONS

Amanda Mueller

Acer negundo
Boxelder

Look for: Twigs shiny green; never hairy. Opposite. Grows as small/medium tree. Sometimes > 3 leaflets. p 23.

Toxicodendron radicans
Eastern Poison Ivy

Look for: Hairy stem/vine. Alternate leaf arrangement. Never grows independently as a small tree or shrub. p 381.

• •

Baccharis halimifolia
Eastern Baccharis

Look for: Wider toothed leaves, alternate, on smooth stems; tufted hairs on flowers. p 97.

Iva frutescens
Marsh Elder

Look for: Succulent, more elongated leaves, opposite, on slightly hairy stems. Rounded flowers. p 131.

Illustrated Comparisons

Clethra alnifolia
Coastal Sweet-pepperbush

Look for: Margins of leaf bases smooth; toothed elsewhere. Straight veins. p 103.

Itea virginica
Virginia Sweetspire

Look for: All margins evenly toothed with tiny spines. Curved veins near margins. p 129.

• •

Eubotrys racemosus
Swamp Fetterbush

Look for: All major veins loop inward toward central vein; margins without spines. Fruit capsules rounded. p 111.

Itea virginica
Virginia Sweetspire

Look for: Margins evenly toothed with tiny spines; lower major veins do not loop inward to touch other major veins. Fruit capsules elongated. p 129.

COMMON CONFUSIONS

Amanda Mueller

Aesculus sylvatica
Painted Buckeye

Look for: Toothed compound leaves; leaflets attached at a central point (palmate). p 89.

Asimina triloba
Common Pawpaw

Look for: Alternate, untoothed leaves, not compound. Leaves malodorous when crushed. p 95.

• •

Onoclea sensibilis
Sensitive Fern

Look for: Leaflets tend to be opposite. Fertile frond is vertical stalk with spherical spore cases. p 169.

Woodwardia areolata
Netted Chain Fern

Look for: Leaflets tend to be alternate; fertile frond is like thin version of non-fertile frond, with elongated spore cases. p 175.

Illustrated Comparisons

Amanda Mueller

Osmundastrum cinnamomeum
Cinnamon Fern

Look for: Absence of chain-like veins along leaflet stalk. Tufts of brown hairs at bases of leaflets (pinnae). p 172.

Woodwardia virginica
Virginia Chain Fern

Look for: Chain-like veins along leaflet stalk. No tufts of hairs at bases of leaflets (pinnae). p 177.

• •

***Iris* spp.**
Iris (Emerging)

Look for: Base is more flattened, fan-like; leaves don't twirl. p 223.

Typha latifolia
Broadleaf Cattail (Emerging)

Look for: Base is more rounded; leaves twirl as emerging. p 275.

COMMON CONFUSIONS

Peltandra virginica
Green Arrow Arum

Look for: Leaves more sharply angled. Leaves have parallel side veins along edges. p 237.

Sagittaria latifolia
Broadleaf Arrowhead

Look for: Leaves more rounded. Major veins all radiate from single point. p 253.

Pontedaria cordata
Pickerelweed

Look for: Heart shaped leaves without points. Veins radiate from base to tip. Purple flowers often present. p 241.

Sagittaria latifolia
Broadleaf Arrowhead

Look for: Leaves with three pointed ends. Veins radiate in two directions from central point. White and yellow flowers. p 253.

Illustrated Comparisons

Phragmites australis
Common Reed

Look for: Finer spikelets generally oriented toward one side of stalk. p 239.

Spartina cynosuroides
Big Cordgrass

Look for: Sturdier spikelets in all directions on stalk. p 269.

• •

Boehmeria cylindrica
False Nettle

Look for: Dull leaves; stems not translucent. Larger perennial with unbranching flowering spikes. p 289.

Pilea pumila
Canadian Clearweed

Look for: Glossy leaves with three prominent, unbranched veins; translucent stems. Smaller size (often < 30 cm). Branching inflorescences. p 329.

COMMON CONFUSIONS

Centella erecta
Erect Centella

Look for: Stem attached to leaf at edge of leaf; veins radiate from leaf bases. p 291.

Hydrocotyle umbellata
Marsh Pennywort

Look for: Stem attached to leaf in center of circular disk; veins radiate from center. p 305.

Campsis radicans
Trumpet Creeper

Look for: Once-divided compound leaves. p 365.

Nekemias arborea
Peppervine

Look for: Twice-divided compound leaves. p 371.

Illustrated Comparisons

Toxicodendron radicans
Eastern Poison Ivy
Look for: Leaves often slightly lobed; woody stem. p 381.

Arisaema triphyllum
Jack-in-the-Pulpit
Look for: No lobes or teeth on leaves. Non-woody stem. p 183.

Nuphar lutea
Yellow Pond-Lily
Look for: Leaves more elongated than water lily, veins parallel. Yellow flower. p 395.

Nymphaea odorata
American Water-lily
Look for: Leaves more circular with more sharply defined cut, branching veins. Fragrant white flower. p 397.

INDEX

Hibiscus moscheutos

A

Acer
- Acer floridanum 25
- Acer negundo 23, 381, 403
- Acer rubrum 25
- Acer saccharum 25
- Acer spp. 65

Aesculus
- Aesculua flava 89
- Aesculus pavia 89
- Aesculus sylvatica 89, 405

Alabama supplejack 361
Alder, tag 91
Allegheny monkey-flower 323
Alligatorweed 283
Alnus serrulata 91
Alternanthera philoxeroides 283
American bur-reed 265
American cupscale 249
American elm 85, 402
American holly 43
American lotus 393
American strawberry-bush 113
American sycamore 25, 65
American water-lily 393, 395, 397, 410

Ampelopsis
- Ampelopsis arborea 365, 371
- Ampelopsis bipinnata 371

Anchistea virginica 177

Andropogon
- Andropogon glaucopsis 181
- Andropogon glomeratus 181

Aneilema keisak 235
Anisostichus capreolata 363
Annual wild-rice 239
Apios americana 359
Aquatic plants 385
Arisaema triphyllum 183, 381, 410
Aristida stricta 185
Aronia arbutifolia 93
Arrow arum, green 237, 241, 253, 407

Arrowhead
- Broadleaf arrowhead 253, 407
- Bull-tongue arrowhead 251

Arrowleaf tearthumb 327

Arrowwood
- Downy arrowwood 163
- Northern arrowwood 163
- Southern arrowwood 163

Arthraxon hispidus 187, 233

Arundinaria
- Arundinaria gigantea 189, 239
- Arundinaria tecta 189

Asclepias incarnata 285

Ash
- Carolina ash 37
- Green ash 39

Asiatic dayflower 201
Asimina triloba 89, 95, 405
Aster pilosus 353
Aster, frost 353
Atamasco lily 279
Atamosco atamasca 279

Athyrium
- Athyrium asplenioides 167
- Athyrium filix-femina 167, 173

Atlantic white cedar 33
Azolla caroliniana 387

B

Baccharis, eastern 97, 131, 403
Baccharis halimifolia 97, 131, 403
Bald cypress 81, 83
Barnyard grass, large 215
Barnyard grass, rough 215

Bay
- Loblolly bay 41
- Swamp bay 49, 57
- Sweetbay 49

Bayberry, southern 145, 147
Beakrush 242
Beaksedge 242
- Angle-stem beaksedge 244
- Clustered beaksedge 245
- Globe beaksedge 245
- Short-bristle horned beaksedge 244

Bedstraw 301
Beggarticks, devil's 287
Berchemia scandens 361
Betula nigra 27
Betula spp. 29, 85, 402

Bidens frondosa 287
Big cordgrass 239, 269, 408
Bignonia capreolata 363, 365
Birch, river 27
Birches 29, 85, 402
Bishopweed, mock 337
Black elderberry 155
Black needlerush 227
Black willow 79, 153
Blackberry, sawtooth 151
Blackgum 35, 53, 55, 139
Bladderwort
    Eastern purple bladderwort 399
    Southern bladderwort 399
    Zigzag bladderwort 399
Blisterwort 341
Blueberry, highbush 159
Blue huckleberry 115
Blue-eyed grass
    Eastern blue-eyed grass 263
    Narrowleaf blue-eyed grass 263
Bluestem
    Bushy bluestem 181
    Purple bluestem 181
Blunt brown sedge 195
Blunt spike-rush 217
Boehmeria cylindrica 289, 329, 408
Bog rush 225
Borrichia frutescens 99
Boxelder 23, 381, 403
Broadleaf arrowhead 253, 407
Broadleaf cattail 273, 275
Buckeye, painted 89, 405
Buckeye, red 89
Bull-tongue arrowhead 251
Bulrush
    Georgia bulrush 261
    Green bulrush 261
    Softstem bulrush 255, 257, 259
    Woodland bulrush 257, 259
    Woolgrass bulrush 255, 257, 259, 261
Bur-reed, American 265
Bursting-heart 113
Bushy bluestem 181
Buttercup, kidney-leaf 339
Buttonbush, common 101, 105, 161

C

Camphorweed, stinking 331
Camphor pluchea 331
Campsis radicans . 363, 365, 371, 383, 409
Canadian clearweed 289, 329, 408
Cane, giant 189, 239
Cardinalflower 315
Carex spp 190
    Carex abscondita 191
    Carex atlantica 191
    Carex blanda 191
    Carex caroliniana 192
    Carex crinita 192
    Carex debilis 192
    Carex glaucescens 193
    Carex grayi 193
    Carex gynandra 192
    Carex intumescens 193
    Carex laevivaginata 193
    Carex lupulina 194
    Carex lurida 194
    Carex oxylepis 194
    Carex scoparia 195
    Carex tribuloides 195
    Carex vulpinoidea 195
Carolina ash 37
Carolina dropseed 185
Carolina jessamine 367
Carolina mosquitofern 387
Carolina sea-lavender 311
Carolina sedge 192
Carpgrass, small 187, 233
Carpinus caroliniana 27, 29, 402
Cattail 273-275, 406
    Broadleaf cattail 273, 275
    Narrowleaf cattail 273, 275
Celtis laevigata 31
Centella
    Centella asiatica 291
    Centella erecta 291, 305, 409
Cephalanthus occidentalis 101, 105, 161
Chain fern
    Netted chain fern 169, 175, 405
    Virginia chain fern 173, 177, 406
Chamaecyparis thyoides 33
Chasmanthium latifolium 197
Cherrybark oak 75

413

Chestnut oak 71
Chinese privet 137
Chokeberry, red 93
Chufa 204
Cicuta maculata 293
Cinnamon fern 173, 406
Cladium
    Cladium jamaicense 199
    Cladium mariscus 199
Clearweed, Canadian 289, 329, 408
Clethra alnifolia 103, 129, 404
Climbing dayflower 201
Climbing hempvine 369
Coastal doghobble 135
Coastal Plain willow 79, 153
Coastal sea-oats 197
Coastal sweet-pepperbush 103, 129, 404
Colombian waxweed 295
Commelina
    Commelina communis 201
    Commelina diffusa 201
    Commelina erecta 201
    Commelina spp. 235
    Commelina virginica 201
Common buttonbush 101, 105, 161
Common fox sedge 195
Common goldstar 221
Common ladyfern 167, 173
Common pawpaw 89, 95
Common persimmon 35, 55
Common reed 239, 247, 269, 408
Common rush 226
Common sweetleaf 157
Common wax myrtle 145, 147
Common water-primrose 319
Common winterberry 123, 127
Common woodland sedge 191
Compressed plumegrass 247
Cordgrass, big 239, 269, 408
Cordgrass, saltmeadow 211, 271
Cordgrass, smooth 267
Cornus
    Cornus amomum 101, 105, 161
    Cornus foemina 105
Creeper, trumpet ... 363, 365, 371, 383, 409
Creeper, Virginia 381, 383
Crossvine 363, 365
Cuphea carthagenensis 295

Cupscale, American 249
Cutgrass, rice 231
Cyperus spp. 202
    Cyperus esculentus 203, 204
    Cyperus flavescens 203, 204
    Cyperus polystachyos 203, 205
    Cyperus pseudovegetus 203, 205
    Cyperus strigosus 203, 205

Cypress
    Bald cypress 81, 83
    Pond cypress 81, 83
Cyrilla racemiflora 107

D

Daisy, false 299
Darlington oak 67
Dayflower 201, 235
    Asiatic dayflower 201
    Climbing dayflower 201
    Virginia dayflower 201
    Whitemouth dayflower 201
Decodon verticillatus 109
Devil's beggarticks 287
Dichanthelium
    Dichanthelium scoparium 209
    Dichanthelium spp. 207, 209
Dicot herbs 281
Diospyros virginiana 35, 55
Distichlis spicata 211
Ditch stonecrop 325
Doghobble, coastal 135
Dogwood, silky 101, 105, 161
Dogwood, swamp 105, 161
Dropseed, Carolina 185
Drosera spp. 297
    Drosera brevifolia 297
    Drosera capillaris 297
    Drosera intermedia 297
    Drosera rotundifolia 297
Duckweed 387, 389
Dulichium arundinaceum 213
Dwarf sundew 297
Dwarf St. John's wort 307

E

Eastern baccharis 97, 131, 403
Eastern blue-eyed grass 263
Eastern hemlock 83
Eastern poison ivy ....
.... 23, 183, 309, 383, 381, 403, 410
Eastern purple bladderwort 399
Eastern red cedar 33
Echinochloa
    Echinochloa crus-galli 215
    Echinochloa muricata 215
Eclipta
    Eclipta alba 299
    Eclipta prostrata 299
Elder, marsh 97, 131, 403
Elderberry, black 155
Eleocharis obtusa 217
Elm
    American elm 85, 402
    Slippery elm 85
    Winged elm 45, 85
Erect centella 291, 305, 409
Erianthus giganteus 247
Eubotrys racemosus 111, 129, 135, 404
Eulalia vimineum 233
Euonymus americanus 113

F

False daisy 299
False nettle 289, 329, 408
False pimpernel, yellow-seed 303, 313
Ferns 165
    Cinnamon fern 173, 177, 406
    Netted chain fern 169, 175, 405
    Royal fern 171
    Sensitive fern 169, 405
    Virginia chain fern 173, 177, 406
Fetterbush lyonia 143
Fetterbush, swamp 111, 129, 135, 404
Fimbristylis spp 219
    Fimbristylis autumnalis 219
    Fimbristylis castanea 219
Fimbry species 219
    Marsh fimbry 219
    Slender fimbry 219

Flatsedge 202
    Manyspike flatsedge 205
    Marsh flatsedge 205
    Straw-color flatsedge 205
    Yellow flatsedge 204
Fleabane, saltmarsh 333
Florida maple 25
Forked rush 226
Fraxinus
    Fraxinus caroliniana 37, 39
    Fraxinus pennsylvanica 37, 39
Fringed sedge 192
Frost aster 353

G

Galium tinctorium 301
Gallberry, large 121
Gaylussacia frondosa 115
Gaylussacia spp 141, 159
Gelsemium sempervirens 367
Georgia bulrush 261
Giant cane 189, 239
Glasswort 345
Globe beaksedge 245
Glossy privet 137
Goldenrod
    Roundleaf goldenrod 349
    Seaside goldenrod 351
Goldstar, common 221
Gordonia lasianthus 41
Grape, graybark 383
Grass
    Large barnyard grass 215
    Rosette grass 207
    Rough barnyard grass 215
    Yellow-eyed grass 217, 277
Grassleaf rush 225, 226
Gratiola virginiana 303, 313
Graybark grape 383
Gray's sedge 193
Greater bladder sedge 193
Green arrow arum 237, 241, 253, 407
Green ash 39
Green bulrush 261
Greenbrier
    Laurel greenbrier 377
    Roundleaf greenbrier 379

Greenbrier (cont'd)
    Saw greenbrier 373, 379
    Whiteleaf greenbrier 375
Groundnut 359
Groundsel tree 97

H

Hearts-a-busting 113
Hedge-hyssop, round-fruit 303, 313
Hemlock, eastern 83
Hemlock, spotted water 293
Hempvine, climbing..................................369
Hibiscus moscheutos 117, 133
Highbush blueberry 159
Holly, American 43
Honeysuckle, Japanese 137, 367
Hop sedge 194
Hornbeam, hop 27, 29
Horse-sugar 157
Huckleberry, blue......................................115
Hydrocotyle umbellata 291, 305, 409
Hypericum
    Hypericum hypericoides 119
    Hypericum mutilum 307
    Hypericum punctatum 307
Hypoxis hirsuta 221

I

Ilex
    Ilex coriacea 121, 125
    Ilex decidua 123, 127
    Ilex glabra 121, 125
    Ilex opaca 43
    Ilex verticillata 123, 127
Impatiens
    Impatiens capensis 309
    Impatiens pallida 309
Indian wood-oats 197
Inkberry...121, 125
Iris spp. 223, 229, 406
Iris virginica 223
Iris, Virginia 223
Ironwood............................... 27, 29, 402
Itea virginica................. 103, 111, 129, 404
Iva frutescens 97, 131, 403

J

Jack-in-the-pulpit 183, 381, 410
Japanese honeysuckle 137, 367
Japanese stilt grass................ 137, 187, 233
Jessamine, Carolina 367
Jewelweed 309
Jewelweed, pale......................................309
Juncus spp..224
    Juncus acuminatus 225
    Juncus biflorus 225
    Juncus coriaceus 224, 225
    Juncus dichotomus 226
    Juncus effusus 224, 226
    Juncus marginatus 225, 226
    Juncus repens..227
    Juncus roemerianus 227
    Juncus scirpoides 227
Juniperus virginiana 33

K

Kalmia latifolia 157
Kidney-leaf buttercup 339
Knotty-leaf rush.......................................225
Kosteletzkya
    Kosteletzkya pentacarpos 133
    Kosteletzkya virginica 117, 133

L

Lachnanthes caroliniana 229
Ladyfern, common....................167, 173
Large barnyard grass 215
Large gallberry 121
Laurel greenbrier 377
Laurel oak... 67, 77
Laurel, mountain.....................................157
Leathery rush 225
Leersia oryzoides 231
Lemna spp................................387, 389
Lesser creeping rush................................227
Leucothoe
    Leucothoe axillaris 135
    Leucothoe racemosa 111

Ligustrum
    Ligustrum lucidum 137
    Ligustrum sinense 137
Limonium carolinianum 311
Lindera benzoin 55, 139
Lindernia dubia 303, 313
Liquidambar styraciflua 45
Liriodendron tulipifera 47
Live oak 67
Lizard's tail 347
Lobelia cardinalis 315
Loblolly bay 41
Loblolly pine 59, 61, 63
Longleaf pine 59, 63
Lonicera japonica 367
Loosestrife, swamp 109
Lorinseria areolata 175
Lotus, American 393
Lowland rotala 319
Ludwigia
    Ludwigia alternifolia 317
    Ludwigia hexapetala 319
    Ludwigia palustris 319
Lycopus virginicus 321
Lyonia
    Lyonia ligustrina 141
    Lyonia lucida 143
Lyonia, fetterbush 143

M

Magnolia virginiana 49
Maleberry 141
Mallow
    Saltmarsh mallow 117, 133
    Swamp rose mallow 117, 133
Manyflower marsh-pennywort 304
Maple
    Florida maple 25
    Red maple 25
    Sugar maple 25
Marsh elder 97, 131, 403
Marsh fimbry 219
Marsh pennywort 291, 305, 409
Marsh primrose-willow 319
Maryland meadowbeauty 343
Meadowbeauty, Maryland 343
Microstegium vimineum 137, 187, 233

Mikania scandens 369
Milkweed, swamp 285
Mimulus spp. 323
    Mimulus alatus 323
    Mimulus ringens 323
Mock bishopweed 337
Monkey-flower
    Allegheny monkey-flower 323
    Sharpwing monkey-flower 323
Monocot herbs 179
Morella
    Morella caroliniensis 145, 147
    Morella cerifera 145, 147
Mosquitofern, Carolina 387
Mountain laurel 157
Multiflora rose 149
Murdannia keisak 235
Myrica
    Myrica cerifera 147
    Myrica heterophylla 145
Myriophyllum spp. 391
    Myriophyllum aquaticum 391
    Myriophyllum heterophyllum 391

N

Narrowleaf blue-eyed grass 263
Narrowleaf cattail 273, 275
Needlepod rush 227
Nekemias arborea 365, 371, 409
Nelumbo lutea 393
Netted chain fern 169, 175, 405
Nettle, false 289, 329, 408
Nodding sedge 192
Northern arrowwood 163
Northern spicebush 55, 139
Nuphar
    Nuphar advena 395
    Nuphar lutea 395, 397, 410
Nutsedge, yellow 204
Nymphaea odorata 393, 395, 397, 410
Nyssa
    Nyssa aquatica 51, 53
    Nyssa biflora 51, 53, 55
    Nyssa sylvatica 35, 53, 55, 139

O

Oak
    Cherrybark oak 75
    Chestnut oak 71
    Darlington oak 67
    Laurel oak 67, 77
    Live oak 67
    Overcup oak 69
    Southern red oak 75
    Swamp chestnut oak 71
    Swamp white oak 71
    Water oak 73, 77
    Willow oak 67, 73, 77
Onoclea sensibilis 169, 175, 405
Osmunda
    Osmunda cinnamomea 173
    Osmunda regalis 171
    Osmunda spectablis 171
Osmundastrum cinnamomeum ....
    .... 173, 177, 406
Ostrya virginiana 27, 29
Overcup oak 69
Ox-eye daisy, sea 99

P

Painted buckeye 89, 405
Pale jewelweed 309
Panicgrass, velvet 209
Parrot feather water-milfoil....................... 391
Parthenocissus quinquefolia 381,383
Pawpaw, common 89, 95, 405
Peltandra virginica 237, 241, 253, 407
Pennywort, marsh.................... 291, 305, 409
Penthorum sedoides 325
Pepperbush, Coastal sweet-... 103, 129, 404
Peppervine 365, 371, 409
Persea
    Persea borbonia 57
    Persea palustris 49, 57
Persicaria spp. 327, 335
    Persicaria sagittata 327
Persimmon, common 35, 55
Photinia pyrifolia 93
Phragmites
    Phragmites australis..239, 247, 269, 408

Phragmites (cont'd)
    Phragmites communis 239
Pickerelweed 241, 253, 347, 407
Pilea pumila 289, 329, 408
Pimpernel, yellow-seed false 303, 313
Pine
    Loblolly pine 59, 61, 63
    Longleaf pine... 59, 63
    Pond pine 61, 63
Pineland threeawn......................................185
Pink sundew 297
Pinus
    Pinus palustris 59, 63
    Pinus serotina 61, 63
    Pinus taeda 59, 61, 63
Platanus occidentalis 25, 65
Pluchea
    Pluchea camphorata 331
    Pluchea foetida 331, 333
    Pluchea odorata 331, 333
Plumegrass
    Compressed plumegrass 247
    Sugarcane plumegrass 239, 247
Poison ivy, eastern...................................
    ....23, 183, 309, 383, 381, 403, 410
Polygonum
    Polygonum sagittatum 327
    Polygonum spp. 335
Pond cypress 81, 83
Pond pine 61, 63
Pond-lily, yellow 395, 397, 410
Pontederia cordata 241, 253, 347, 407
Possumhaw 123, 127
Possumhaw viburnum................................161
Prickly bog sedge 191
Primrose-willow, marsh 319
Privet
    Chinese privet 137
    Glossy privet..137
Ptilimnium capillaceum 337
Purple bladderwort, eastern 399
Purple bluestem..181

Q

Quercus
- *Quercus bicolor* 71
- *Quercus falcata* 75
- *Quercus hemisphaerica* 67
- *Quercus laurifolia* 67, 77
- *Quercus lyrata* 69
- *Quercus michauxii* 71
- *Quercus montana* 71
- *Quercus nigra* 73, 77
- *Quercus pagoda* 75
- *Quercus phellos* 67, 73, 77
- *Quercus virginiana* 67

R

Ranunculus
- *Ranunculus abortivus* 339
- *Ranunculus recurvatus* 341

Red buckeye 89
Red cedar, eastern 33
Red chokeberry 93
Red maple 25
Red oak, southern 75
Redbay, upland 57
Redroot 229
Reed, common 239, 247, 269, 408
Rhexia mariana 343
Rhynchospora spp. 242
- *Rhynchospora caduca* 243, 244
- *Rhynchospora corniculata* 243, 244
- *Rhynchospora globularis* 243, 245
- *Rhynchospora glomerata* 243, 245
- *Rhynchospora recognita* 245

Rice cutgrass 231
River birch 27
Rosa multiflora 149
Rosa palustris 149
Rose
- Multiflora rose 149
- Swamp rose 149

Rosette grass 207
Rotala ramosior 319
Rough barnyard grass 215
Round-fruit hedge-hyssop 303, 313
Roundleaf goldenrod 349
Roundleaf greenbrier 379
Roundleaf sundew 297
Royal fern 171
Rubus
- *Rubus argutus* 151
- *Rubus pensilvanicus* 151

Rushes 224
- Black needlerush 227
- Bog rush 225
- Common rush 226
- Forked rush 226
- Grassleaf rush 225, 226
- Knotty-leaf rush 225
- Leathery rush 225
- Lesser creeping rush 227
- Needlepod rush 227

S

Saccharum
- *Saccharum coarctatum* 247
- *Saccharum giganteum* 239, 247

Sacciolepis striata 249
Sagittaria
- *Sagittaria falcata* 251
- *Sagittaria lancifolia* 251
- *Sagittaria latifolia* 237, 253, 407

Salicornia spp. 345
- *Salicornia ambigua* 345
- *Salicornia bigelovii* 345
- *Salicornia virginica* 345

Salix
- *Salix caroliniana* 79, 153
- *Salix nigra* 79, 153

Saltgrass 211
Saltmarsh fleabane 331
Saltmarsh mallow 117, 133
Saltmeadow cordgrass 211, 271
Sambucus
- *Sambucus canadensis* 155
- *Sambucus nigra* 155

Saururus cernuus 347
Saw greenbrier 373, 379
Sawgrass, swamp 199
Sawtooth blackberry 151
Schoenoplectus tabernaemontani ....
.... 255, 257, 259

Scirpus
   Scirpus atrovirens 261
   Scirpus cyperinus 255, 257, 259, 261
   Scirpus expansus 257, 259
   Scirpus georgianus 261
   Scirpus validus................................... 255
Sea oats... 197
Sea ox-eye daisy 99
Sea-lavender, Carolina............................ 311
Seaside goldenrod................................... 351
Sedges
   Blunt brown sedge................................ 195
   Broom sedge 195
   Carolina sedge 192
   Common fox sedge............................... 195
   Common woodland sedge 191
   Fringed sedge 192
   Gray's sedge 193
   Greater bladder sedge 193
   Hop sedge 194
   Nodding sedge...................................... 192
   Prickly bog sedge.................................. 191
   Shallow sedge....................................... 194
   Sharpscale sedge 194
   Smooth-sheathed sedge........................ 193
   Southern waxy sedge 193
   Thicket sedge.. 191
   White-edge sedge 192
Seedbox.. 317
Sensitive fern 169, 175, 405
Shallow sedge 194
Sharpscale sedge 194
Sharpwing monkey-flower 323
Shrubs 87
Silky dogwood 101, 105, 161
Sisyrinchium
   Sisyrinchium angustifolium 263
   Sisyrinchium atlanticum 263
Slender fimbry....................................... 219
Slippery elm 85
Small carpgrass...............................187, 233
Smartweed 335
Smilax
   Smilax bona-nox 373, 379
   Smilax glauca..................................... 375
   Smilax laurifolia 377
   Smilax rotundifolia 379
Smooth cordgrass 267
Smooth-sheathed sedge......................... 193

Softstem bulrush 255, 257, 259
Solidago
   Solidago mexicana 351
   Solidago patula 349
   Solidago sempervirens 351
Southern bayberry 145, 147
Southern bladderwort 399
Southern red oak 75
Southern waxy sedge 193
Sparganium americanum..................... 265
Spartina
   Spartina alterniflora 267
   Spartina cynosuroides 239, 269, 408
   Spartina patens 211, 271
Spicebush, northern 55, 139
Spoonleaf sundew 297
Sporobolus
   Sporobolus alterniflorus 267
   Sporobolus cynosuroides 269
   Sporobolus pinetorum 185
   Sporobolus pumilus............................ 271
Spotted St. John's wort 307
Spotted water hemlock 293
St. Andrew's-cross.................................. 119
St. John's wort
   Dwarf St. John's wort............................ 307
   Spotted St. John's wort 307
Stilt grass, Japanese 137, 187, 233
Stinking camphorweed 331
Stonecrop, ditch..................................... 325
Strawberry-bush, American 113
Sugar maple 25
Sugarberry 31
Sugarcane plumegrass 239, 247
Sundew species 297
   Dwarf sundew 297
   Pink sundew... 297
   Roundleaf sundew................................ 297
   Spoonleaf sundew 297
Supplejack, Alabama.............................. 361
Swamp bay 49, 57
Swamp chestnut oak................................ 71
Swamp dogwood.................................... 105
Swamp fetterbush 111, 129, 135, 404
Swamp loosestrife 109
Swamp milkweed................................... 285
Swamp rose 149
Swamp rose mallow 117, 133
Swamp sawgrass.................................... 199

Swamp titi 107
Swamp tupelo 51, 53, 55
Swamp white oak 71
Sweetbay 49
Sweetgum 45
Sweetleaf, common 157
Sweet-pepperbush, coastal 103, 129, 404
Sweetscent 333
Sweetspire, Virginia 103, 111, 129, 404
Sycamore, American 25, 65
Symphyotrichum
    Symphyotrichum dumosum 353
    Symphyotrichum pilosum 353
Symplocos tinctoria 157

T

Tag alder 91
Taxodium
    Taxodium ascendens 81, 83
    Taxodium distichum 81, 83
Tearthumb, arrowleaf 327
Threeawn, pineland 185
Three-way sedge 213
Thicket sedge 191
Titi, swamp 107
Toxicodendron radicans ....
    .... 23, 183, 309, 383, 381, 403, 410
Trees 21
Trumpet creeper 363, 365, 371, 383, 409
Tsuga canadensis 83
Tuliptree 47
Tupelo
    Swamp tupelo 51, 53, 55
    Water tupelo 51, 53
Twoleaf water-milfoil 391
Typha
    Typha angustifolia 273, 275
    Typha latifolia 273, 275, 406
    Typha spp. 223, 273, 275

U

Ulmus
    Ulmus alata 45, 85
    Ulmus americana 85, 402
    Ulmus rubra 85

Uniola paniculata 197
Upland redbay 57

Utricularia spp. 399
    Utricularia juncea 399
    Utricularia purpurea 399
    Utricularia subulata 39

V

Vaccinium corymbosum 159
Vaccinium spp. 115, 141
Velvet panicgrass 209
Verbena urticifolia 355
Vervain, white 355
Viburnum
    Viburnum dentatum 163
    Viburnum nudum 161
    Viburnum rafinesquianum 163
    Viburnum recognitum 163
Viburnum, possumhaw 161
Vines 357
Virginia chain fern 173, 177, 406
Virginia creeper 381, 383
Virginia dayflower 201
Virginia iris 223
Virginia sweetspire 103, 111, 129, 404
Virginia water horehound 321
Vitis
    Vitis baileyana 383
    Vitis cinerea 383
    Vitis simpsonii 383

W

Wart-removing-herb 235
Water hemlock, spotted 293
Water horehound, Virginia 321
Water oak 73, 77
Water tupelo 51, 53
Water-lily, American 393, 395, 397, 410
Water-milfoil
    Parrot feather water-milfoil 391
    Twoleaf water-milfoil 391
Water-primrose, common 319
Water-willow 109
Wax myrtle, common 145, 147

Waxweed, Columbian 295
White cedar, Atlantic 33
White-edge sedge 192
White vervain 355
Whiteleaf greenbrier 375
Whitemouth dayflower 201
Wild-rice, annual 239
Willow
    Black willow 79, 153
    Coastal Plain willow 79, 153
    Water-willow 109
Willow oak 67, 73, 77
Willow, black 79
Winged elm 45, 85
Winterberry, common 123, 127
Woodland bulrush 257, 259
Wood-oats, Indian 197
Woodwardia
    Woodwardia areolata 169, 175, 405
    Woodwardia virginica 173, 177, 406
Woolgrass bulrush 255, 257, 259, 261

X

Xyris spp. 217, 277

Y

Yellow buckeye 89
Yellow pond-lily .... .395, 397, 410
Yellow-eyed grass 217, 277
Yellow-seed false pimpernel 303, 313

Z

Zephyranthes
    Zephyranthes atamasca 279
    Zephyranthes atamasco 279
Zigzag bladderwort 399
Zizania aquatica 239

Made in the USA
Middletown, DE
14 July 2023